BIOCHEMISTRY RESEARCH TRENDS

ORGANIC ACIDS

CHARACTERISTICS, PROPERTIES AND SYNTHESIS

BIOCHEMISTRY RESEARCH TRENDS

Additional books in this series can be found on Nova's website under the Series tab.

Additional e-books in this series can be found on Nova's website under the e-book tab.

BIOCHEMISTRY RESEARCH TRENDS

ORGANIC ACIDS

CHARACTERISTICS, PROPERTIES AND SYNTHESIS

CESAR VARGAS
EDITOR

nova publishers
New York

Library of Congress Cataloging-in-Publication Data

Names: Vargas, Cesar (Chemistry) editor.
Title: Organic acids : characteristics, properties and synthesis / editor, Cesar Vargas.
Description: Hauppauge, New York: Nova Science Publisher's, Inc., [2016] |
Series: Biochemistry research trends | Includes bibliographical references and index.
Identifiers: LCCN 2016034022 (print) | LCCN 2016039671 (ebook) | ISBN 9781634859318 (hardcover) | ISBN 9781634859523 (ebook) | ISBN 9781634859523
Subjects: LCSH: Organic compounds. | Organic compounds--Synthesis.
Classification: LCC QD271 .O68 2016 (print) | LCC QD271 (ebook) | DDC 547/.037--dc23
LC record available at https://lccn.loc.gov/2016034022

Published by Nova Science Publishers, Inc. † New York

CONTENTS

PREFACE

Organic acids are compounds with acidic properties and occur naturally in a number of foods. They are mainly present in fermented products as a result of hydrolysis, biochemical metabolism, and microbial activity. This book provides research on the characteristics, properties and synthesis of organic acids. Chapter One reviews the application of organic acids in food preservation. Chapter Two provides a chromatographic analysis organic acids in food from animal origin. Chapter Three discusses the diversity of organic acids content in green and roasted *Coffea arabica* cultivars. Chapter Four studies the effects of organic acids and polysaccharides on the solubility of oyster-derived zinc digested *in vitro*. Chapter Five examines the evaporation of organic acids aqueous solutions through spread films of polyelectrolyte/surfactant complexes. Chapter Six discusses the influence of liquid absorption and the effect of downstream pressure on the separation performance of pervaporation of acetic acid aqueous solutions.

Chapter 1 - Increasing world population beyond the present 7 billion and the devastating effect of climate change require newer advanced technology to make wholesome food available and at the right amount. The use of various types of preservatives of both natural and anthropogenic origin has found application in food industries. Recurrent reports of food poisoning (due to the use of chemical preservatives) among other reasons have led to the search for safe and effective preservatives mostly of plant origin. Organic acids (OAs) have therefore been used as an effective natural intervention to reduce spoilage of food products. They are described as low-molecular weight carbohydrate containing compounds which are found in all organisms and characterised by the possession of one or more carboxyl groups. The most documented OAs used as food preservatives include acetic, citric, formic, lactic, propionic, sorbic and benzoic acid. Mechanism of inactivation by these acids is the ability of un-dissociated form to penetrate through the cell membrane, dissociate inside the cell, resulting in decreased intracellular pH value, which is essential for the control of ATP synthesis, RNA and protein synthesis, DNA replication and cell growth. Commercial status of OAs has long been approved based on concentration in various uses such as de-contamination. The chapter features OAs in foods, food products naturally containing OAs, their groups, combinations and structural description, synthesis and inhibition in microorganisims, large-scale industrial production, application and roles. It is expected that knowledge of OAs will increase its application in food preservation processes thereby assuring safe and quality foods that is free from unacceptable risk and hazards.

Chapter 2 - Organic acids are compounds with acidic properties and occur naturally in a number of foods. They are mainly present in fermented products as a result of hydrolysis, biochemical metabolism, and microbial activity. Even so these are not considered as nutrients, however, they are responsible for giving a characteristic taste to food. In addition, the organic acids have been widely used as food additives and preservatives for avoiding food deterioration and extending the shelf life of different products. For these reasons, determining organic acid content in food products is important, since these compounds contribute to the flavor and aromatic properties of them. Besides, organic acids can influence the preservation of some foods. However, they are not members of a homologous series, which differ in the number of carboxy groups, hydroxy groups, and carbon–carbon double bonds in their molecules. The lowest monocarboxylic aliphatic acids, such as formic, acetic, propionic, and butyric, are rather volatile liquids, whereas those acids containing more carbon atoms are of a relatively oily substance and slightly water soluble. Alicyclic acids are less water soluble than the previous ones. In comparison, dicarboxylic acids are colorless solids with melting points at about 100°C. All these acids form somewhat soluble metal salts and esters, of which the latter are adequately volatile for gas chromatography (GC) analysis using flame ionization or mass spectrometric (MS) detectors. Besides that, they also have spectral absorption properties that make them suitable for high-performance liquid chromatography (HPLC) analysis using ultraviolet, refractive index or MS detectors. In this context, GC has been used to determine the volatile organic acids, while HPLC has been widely used for analyzing non-volatile organic acids in different matrixes. There are some studies evaluating organic acids in honey, dairy, fish and meat products, however, the best tool to each food matrix depend on of the ingredients and the technological process applied by the food industry.

Chapter 3 - Organic acids in coffees are influenced by several factors. The aim of this research was to analyze the profile of organic acids (quinic, malic, citric, acetic, and lactic) and chlorogenic acids in green and roasted coffees cultivar and pH and titrable acidity (TA) in brew coffee. Sixteen cultivars (traditional and modern) grown and harvested in the same place were evaluated. Hierarchical cluster analysis of green coffees formed three groups. G1-Bourbon, Catuaí, Icatu and Catuaí SH2SH3 derived cultivars were associated to higher malic, citric and intermediate value of quinic and 5-CQA. G2 (Catuaí, Icatu) containing higher level of quinic, citric acid and lower value of malic and 5-CQA and G3 (Sarchimor derived) presented lower values of quinic, citric and higher content of 5-CQA. Roasted coffee and brews characteristics formed three groups. G1 (IPR 99, Icatu and Catuaí SH2SH3 derived) with higher values of quinic, pH and intermediate acetic acid values and lower malic, citric, 5-CQA, lactic, and TA. G2 (Bourbon, IPR 97) presented higher content of malic, latic, 5-CQA acids and acidity and intermediate of citric and lower of quinic, acetic acids and pH. G3 (Catuaí, Icatu, Sarchimor, Catuaí and Icatu derived) showed higher value of quinic, acetic, citric, and intermediate values of 5-CQA, lactic, malic acids, pH and TA. The authors can verify that the TA did not differ between the groups but different acids contributed for the formation of TA. It was verified that roasting process provides different acids profiles associated to genetic origin of cultivars.

Chapter 4 - Zinc is an essential element for the human body as it plays a variety of physiological roles. For example, zinc enhances apoptosis and is necessary to maintain the structure of proteins such as zinc finger proteins. Furthermore, zinc deficiency causes disturbances of growth, taste disorders, and hypogonadism issues. Most dietary zinc is absorbed in the small intestine and its bioavailability depends on the components coexisting in the digested food. The effect of the presence of organic acids and polysaccharides on the solubility of oyster-derived zinc (*Crassostrea gigas*) during *in vitro* digestion was examined using pepsin. The concentration of soluble zinc slightly decreased upon addition of organic acids (e.g., citric, malic, sorbic, or lactic) and remained nearly constant while varying the organic acid to zinc molar ratio. Phytic acid significantly lowered the concentration of soluble zinc, with a negative correlation between the liberated zinc and the added phytic acid. The solubility of zinc during digestion can be affected by organic acid-induced chelation and insolubilization processes. The concentration of soluble zinc slightly decreased in the presence of starch or cellulose. This concentration was independent of the amount of polysaccharide. On the contrary, the concentration of soluble zinc decreased with the added amount of alginic acid, pectin, or chitosan. The content of acidic and basic amino acids (e.g., aspartic acid, glutamic acid, lysine, and histidine) in the oyster protein was *ca.* 30% by mole. Therefore, the electrostatic interaction between an electrolytic polysaccharide and the oyster protein might suppress the solubilization of digested oyster-derived zinc.

Chapter 5 - Wilhelmy plate technique was used to determine surface tension isotherms of aqueous solutions of formic and acetic acids with concentration up to 30 vol% and with a film of solution of polyelectrolyte/surfactant complexes spread to the surface. Complexes of sodium polystyrene sulfonate / dodecyl trimethyl ammonium bromide and poly (diallyldimethylammonium chloride)/ sodium dodecylsulfate were used. The formation conditions of films with the extreme concentration of the complexes were determined. The films reduce the evaporation rate of acids solutions by 3-6% and demonstrate selective properties increasing the water content in the vapor by 1-5 abs%. Evaporation tests carried out by means of helium blowing over surface of the liquid with spread film.

Chapter 6 - Pervaporation is a more energy saving, *environmentally* safe and *clean* technology of liquid mixture separation as compared with the existing techniques such as distillation. At present, the effective separation of aqueous organic solutions by using pervaporation is one of the actual tasks of membrane technology. The effectiveness of liquid separation considerably depends on the conditions of the experiments and sorption properties respect to liquid mixtures.

A manuscript presented the influence of sorption capacity and downstream pressure on pervaporation results. Acetic acid aqueous solution were used as model separation mixtures. The main sorption characteristics (equilibrium swelling and composition of sorbates) with respect to various compositions of feed mixtures were studied. Pervaporation experiments were performed in the range of 1 - 50 mm Hg of downstream pressure using the membranes based on cellulose hydrate. It was shown that membranes exhibit higher dehydrating properties with increase water concentration in vapor. The opposite trend - decrease of pervaporation flux and increase of separation factor with the increases of downstream pressure was observed. It was found that contribution of swelled membrane layer to the values of selectivity was significant and achieved up to 60% at downstream pressure of 30 mm Hg and higher.

In: Organic Acids
Editor: Cesar Vargas
ISBN: 978-1-63485-931-8

Chapter 1

APPLICATION OF ORGANIC ACIDS IN FOOD PRESERVATION

T. A. Anyasi[1,*], A. I. O. Jideani[1], J. N. Edokpayi[2] and C. P. Anokwuru [3]

[1]Department of Food Science and Technology, School of Agriculture, University of Venda, Limpopo Province, South Africa
[2]Department of Hydrology and Water Resources, School of Environmental Sciences, University of Venda, Limpopo Province, South Africa
[3]Department of Chemistry, School of Mathematical and Natural Sciences, University of Venda, Limpopo Province, South Africa

ABSTRACT

Increasing world population beyond the present 7 billion and the devastating effect of climate change require newer advanced technology to make wholesome food available and at the right amount. The use of various types of preservatives of both natural and anthropogenic origin has found application in food industries. Recurrent reports of food poisoning (due to the use of chemical preservatives) among other reasons have led to the search for safe and effective preservatives mostly of plant origin. Organic acids (OAs) have therefore been used as an effective natural intervention to reduce spoilage of food products. They are

* Corresponding author Email: tonna.anyasi@univen.ac.za; tonna.anyasi@gmail.com.

described as low-molecular weight carbohydrate containing compounds which are found in all organisms and characterised by the possession of one or more carboxyl groups. The most documented OAs used as food preservatives include acetic, citric, formic, lactic, propionic, sorbic and benzoic acid. Mechanism of inactivation by these acids is the ability of un-dissociated form to penetrate through the cell membrane, dissociate inside the cell, resulting in decreased intracellular pH value, which is essential for the control of ATP synthesis, RNA and protein synthesis, DNA replication and cell growth. Commercial status of OAs has long been approved based on concentration in various uses such as de-contamination. The chapter features OAs in foods, food products naturally containing OAs, their groups, combinations and structural description, synthesis and inhibition in microorganisims, large-scale industrial production, application and roles. It is expected that knowledge of OAs will increase its application in food preservation processes thereby assuring safe and quality foods that is free from unacceptable risk and hazards.

Keywords: organic compounds, preservatives, additives, antimicrobial, spoilage

1. Introduction

For a very long time, food storage has always been at odds with food spoilage. Some of the earliest evidence of food preservation dates back to the post-glacial era, from 15,000 to 10,000 BC. Biological methods was first used from 6000 to 1000 BC when fermentation was used to produce beer, bread, wine, vinegar, yoghurt, cheese and butter (Soomro et al., 2002). Louis Pasteur in 1864 proved that microorganisms in foods were the cause of food spoilage and that these microorganisms were destroyed by heat treatment. Subsequently, a major development in the distribution and storage of foods was introduced in 1940 with the availability of low cost home refrigerators and freezers. This was also followed by the developments of other storage methods such as artificial drying, vacuum packaging, ionizing radiations and chemical preservation. Presently, consumers are exhibiting growing concerns about the use of synthetic chemicals used as preservatives in food with the resultant trend towards less processed food. These unprocessed foods can serve as a habitat for hazardous pathogens which can multiply under refrigerated and anaerobic conditions. Thus a solution to this predicament is the employment of antimicrobial metabolites of fermentative microorganisms in the preservation of these food produce (Soomro et al., 2002).

One of the most important interventions for the control of microbial growth and the overall safety of food products is the use of OAs and other chemical treatment for food preservation. Decontamination of food products using chemical treatments has been one of the most important interventions for controlling their microbiological safety and quality. This is due to the fact that in foods, OAs constitute an inexpensive and effective means of reducing the prevalence and population size of microorganisms. This enables their frequent use in decontamination applications in food commodities. (Loretz et al., 2010, 2011a, 2011b; Ölmez and Kretzschmar, 2009; Rajkovic et al., 2010). Aiming at reducing the prevalence and populations of bacterial contaminants, decontamination of foods results in an improvement in their microbiological status (Hugas and Tsigarida, 2008). Over the years, several decontamination technologies have been established and found to be effective against microbial contaminants of foods with such technologies applied on raw materials, during processing and on the final products. Such technologies which can be used in isolation or in combination includes hot water treatment, steam vacuuming, OAs, use of chlorine, bacteriophages and bacteriocins (Hugas and Tsigarida, 2008; Koutsoumanis et al., 2004; Lianou et al., 2012).

OAs are a group of natural compounds known as weak acids used as food additives, but not all of them have antimicrobial activity. The most effective antimicrobials are acetic acid, lactic acid, propionic acid, sorbic acid and benzoic acid. The activity of OAs is related to pH and to the undissociated form of the acid. The use of OAs is generally limited to foods with a pH less than 5.5 (Doores, 1993). Another factor affecting the activity and use of organic acid (OA) is the polarity. This relates both to the ionization of the molecule and to the contribution of any alkyl side groups or hydrophobic parent molecules. Antimicrobials must be lipophilic to attach and pass through the cell membrane but also soluble in the aqueous phase (Davidson, 1997). The mechanism of action of OAs and their esters have some common elements. In the undissociated form, OAs can penetrate the cell membrane lipid bilayer more easily. Once inside the cell, the acid dissociates because the cell interior has a higher pH than the exterior (Mastromatteo et al., 2010). Bacteria maintain internal pH near neutrality to prevent conformational changes to the cell structural proteins, enzymes, nucleic acids and phospholipids. Protons generated from intracellular dissociation of OA, acidify the cytoplasm and are extruded to the exterior (Davidson and Sofos, 2005;

Mastromatteo et al. 2010). Some of the most important and frequently used additives in beverage, food, and feed production are OAs and their derivatives.

OAs are organic compounds with acidic properties. They are known as weak acids because they dissociate partially in aqueous solution. Generally, the acidity of an OA is determined mainly by the relative stability of the conjugate base of the molecule. They act as buffers when in aqueous solution. OAs differ in the number of the carboxy groups, hydroxy groups and carbon–carbon double bonds in their molecules (Theron and Lues, 2011). They can be classified based on the following criteria: the type of carbon chain that is: aliphatic, alicyclic, aromatic, and heterocyclic; being saturated or unsaturated; being substituted or non-substituted; the number of functional groups. For example, acetic acid (one carboxyl group), malic acid (two carboxyl groups), and citric acid (three carboxyl groups) (Theron and Lues, 2011). They exist as pure acids or in their salt forms (e.g., benzoic acid and sodium benzoate). The most common OAs are the carboxylic acids, with acidity associated with their carboxylic group –COOH (Ricke, 2003; Theron and Lues, 2011; Quitmann et al., 2014).

Table 1. Organic acids commonly used in foods

Organic Acid	Molecular formula	Molecular weight(g/mol)	Appearance	Melting point	Boiling point (°C)
Acetic	$C_2H_4O_2$	60.1	Colourless liquid	16.6	118.2
Benzoic	$C_7H_6O_2$	122.1	Colourless crystalline solid	122.4	250.0
Citric	$C_6H_8O_7$	192.1 (anhydrous) 210.1 (monohydrate)	White crystalline solid	153 (anhydrous)	175 (decomposes)
Formic	CH_2O_2	46.0	Colourless liquid	8.4	100.8
Lactic	$C_3H_6O_3$	90.1	L: hygroscopic crystalline D: hygroscopic plates DL: hygroscopic crystalline or syrupy	L: 53 D: 53 DL: 16.8	122
Propionic	$C_3H_6O_2$	74.1	Colourless liquid	-21	141

Source: Lianou et al. (2012).

Although the lower molecular weight acids such as formic and acetic acids are water soluble, the high molecular weight compounds are insoluble due to increase in the alkyl groups present. However, most OAs are very soluble in organic solvents (Theron and Lues, 2011; Lianou et al., 2012). The OAs most commonly used in foods along with their basic physical and chemical properties are presented in Table 1.

OAs are either naturally present in foods or chemically synthesized and added, directly or indirectly, to food products, with some of them formed during fermentation of carbohydrates in foods (Stratford and Eklund, 2003; Theron and Lues, 2011). OAs have been exploited for a long time as food additives and preservatives in use against food deterioration (Ricke, 2003). Some of these OAs such as acetic, benzoic, citric, formic, lactic, and propionic acid, are already known chemical preservative agents exhibiting a broad spectrum of antimicrobial and enzymatic activities (Brul and Coote, 1999; Theron and Lues, 2007; Anyasi et al., 2015). Acetic and lactic acids, in particular, are commonly used as cheap, environmentally friendly and effective interventions for reducing the levels and prevalence of bacterial pathogens in food products (Rajkovic et al., 2010; Siragusa, 1995). As a result of their bactericidal properties, OAs have constituted the basic active ingredients of antimicrobial products developed and evaluated for the reduction of foodborne pathogenic bacteria on food surfaces (Gonzalez et al., 2004; Laury et al., 2009; López-Gálvez et al., 2009; Okolocha and Ellerbroek, 2005; Lianou et al., 2012).

Chemical decontamination interventions, aiming at improving the microbiological safety and quality of food, are usually implemented at the post-harvest level and at different points of the food processing chain. The efficacy of decontamination treatments, using OAs or other chemical agents, will depend on the type of food and food product being treated, its initial microbial load and ecology, the type of bacterial contaminants to be inactivated, as well as the ability of bacterial contaminants to attach to the treated food and form biofilms (Chaiyakosa et al., 2007; Hugas and Tsigarida, 2008; Kim and Marshall, 2000a, 2001, 2002; Rajkovic et al., 2010). According to literature, the longer the time interval between contamination and decontamination and the higher the temperature encountered during this interval, the more difficult it is to decontaminate fresh produce commodities due to firmer attachment of bacterial cells (Ells and Hansen, 2006; Kondo et al., 2006; Ölmez, 2010; Sapers, 2001; Singh et al., 2002; Ukuku and Sapers, 2001; Ukuku et al., 2001; Lianou et al., 2012).

Various OAs have been investigated for their decontamination efficacy when applied in miscellaneous food products with lots of data reported in several literatures. Some of the research findings can be attributed to differences associated with the decontamination technology applied, as well as other parameters that may also affect the decontamination efficacy of the OAs. Some of such parameters include the nature of acid, the type of treated tissue in the food produce, its pH and buffering capacity, the targeted microorganism as well as the initial microbial population (Rajkovic et al., 2010; Riedel et al., 2009). With the p*K*a of an acid representing the pH at which equal proportions of dissociated and undissociated molecules are present in solution, the antimicrobial activity of OAs increases as the environmental pH approaches the p*K*a (Stratford and Eklund, 2003). Hence, the antimicrobial activity of a certain OA relative to that exerted by another, depends on the buffering capacity of the food matrix and the quantity of the acidulant used (Lianou et al., 2012).

Several authors have suggested that the combined application of several OAs in use as preservatives is an effective decontamination intervention in various foods including meat and fresh produce (Anderson et al., 1992; Dubal et al., 2004; Ölmez, 2010; Podolak et al., 1996). The authors opined that the combination of OAs led to an observed synergistic interaction with regards to their antimicrobial activity. For instance, the synergistic effects observed on mixing lactic and acetic acids on weakly buffered media, have been credited to the potentiation of acetic acid at the lower pH environment created by lactic acid (Helander et al., 1997). Similarly, the natural presence of acetic acid and citric acid in vinegar and lemon juice, offers the opportunity for their application as food decontamination agents throughout the food chain. Solutions of 'household sanitizers' such as vinegar and lemon juice hold promise as decontamination treatments for commodities such as carrots, lettuce and parsley (Chang and Fang, 2007; Karapinar and Gönül, 1992; Sengum and Karapinar, 2004; Vijayakumar and Wolf-Hall, 2002; Wu et al., 2000). The combination of OAs has also been associated with prolonged bacteriostasis during storage of food products resulting in significant shelf life extension (Dubal et al., 2004; Goddard et al., 1996; Marshall and Kim, 1996). The bacteriostatic effect exerted by OAs and their mixtures is even more important in for instance the processed meat products, due to its shelf life extension effect as well as the ready-to-eat (RTE) status of many of these products (Lianou et al., 2012).

2. Organic Acids Groups and Their Application in Food Preservation

OAs are grouped into the monocarboxylic, dicarboxylic, alpha hydroxyl and sugar acids. Monocarboxylic acids include formic, acetic, propionic and sorbic acid. According to literature, acetic acid is used as emulsifiers, stabilizers, preservatives, flavour enhancers and firming agents (Smith and Hong-Shun, 2011). Formic acid which is the simplest carboxylic acid with one carbon atom, is used as a preservative, acidifier in animal feed as well as a flavouring agent at low concentrations (Burdock, 2004). Propionic acid can be used as a pH control agent, preservative and flavour enhancer while sorbic acid has found application as preservatives (Smith and Hong-Shun, 2011; Quitmann et al., 2014).

Dicarboxylic acid includes adipic, fumaric and succinic acid and have found application in beverage, feed and food preservation. Salts of adipic acid such as calcium and magnesium adipate are used in food processes as sequestrants, acidity regulators, baking additives, preservatives and flavour enhancers. Fumaric acid can be used as pH control agents, flavour enhancers, firming agents and as emulsifiers and dough conditioner during esterification process. Succinic acid is also used as flavour enhancers, preservatives, pH control agents and in baking (Quitmann et al., 2014).

Alpha hydroxyl acids which include citric, lactic and malic acid have found application in beverage, food and animal nutrition. Citric acid and its salts are used as sequestrants, pH regulators, preservatives, flavour enhancers, and firming agents. Esters of citric acid can also be used as emulsifiers and solvents. Lactic acid has been used for a long time as acidity regulators, preservatives, baking additives, and flavor enhancers (Anyasi et al., 2015). Lactic acid can also be used as a humectant due to its hygroscopic activity. Malic acid on its part can be used as synergists, acidity regulators, preservatives, and flavour enhancers (Quitmann et al., 2014).

Other OAs are called the sugar acid and they include ascorbic, gluconic, lactobionic and tartaric acid. Ascorbic acid and its isomer erythorbic acid are used as antioxidants, synergist, sequestrants and reducing agents (Smith and Hong-Shun, 2011). Gluconic acid and its salt are used as processing aids in the prevention of milkstone in dairy industry and also in animal nutrition. Lactobionic acid is used as gelling agent, flavour enhancer, antioxidant, acidity regulators, baking additives and firming agents (Quitmann et al., 2014).

2.1. Organic Acid Groups

2.1.1. Monocarboxilic Groups

The chemistry and antimicrobial activity of the saturated straight-chain monocarboxylic acids have been documented as well as derivatives of this group— for example, unsaturated (cinnamic, sorbic), hydroxylic (citric, lactic), phenolic (benzoic, cinnamic, salicylic) and multicarboxylic (azelaic, citric, succinic) acids (Cherrington, et al., 1991). OAs are distinguished from other acids by the carboxylic functional group -COOH to which an OA group or a hydrogen atom are attached. Common names used to describe this group of organic compounds include fatty, volatile fatty, lipophilic, weak, or carboxylic acids (Cherrington, et al., 1991). The acetic acid of vinegar, the formic acid of red ants, and the citric acid of fruits all belong to the same family of compounds - carboxylic acids (Anon 2012). Table 2 shows the position of acetic acid among straight-chain, saturated carboxylic acids. Aspects of the use of OAs include but not limited to animal husbandry as animal feed additives and in abattoirs and food-processing plants where they may be used in controlling microbial contamination of carcass meat (Cherrington et al., 1991).

2.1.1.1. Uses, Applications and Commercial Status

Traditionally, the principal use of acetic acid is in food and food-related applications, while other uses are for non-food industrial applications (https://en.wikipedia.org/wiki/Acetic_acid). As a weak acid, is frequently used as an inexpensive and effective intervention to reduce number and prevalence of bacterial pathogens on food products. Of all OAs evaluated in literature, acetic and lactic acid are found to be the most acceptable (Yasothai and Giriprasad, 2015a). At the same pH, OAs have a greater taste effect than inorganic acids (such as hydrochloric acid). The most common acids found in foods and compared with hydrochloric acid are presented in Table 3 (deMan, 1999).

The application of 2% lactic acid spray solution on beef carcasses and chicken breasts has been effective in reducing population of *E. coli* O157:H7 for more than 1.5 log CFU/cm^2. OAs such as acetic, lactic and citric acids at concentration of 1.5–2.5% have been approved as acceptable innervations for reduction of microbial pathogens on meat carcasses in the United States (Yasothai and Giriprasad, 2015a). European Union provided the legal bases to permit the use of substances other than potable, clean water to decontaminate products of animal origin (EU, 2004).

Table 2. Position of acetic acid among straight-chain, saturated carboxylic acids

Carbon atoms	Common name	IUPAC name	Chemical formula	Common location or use
1	Formic acid	Methanoic acid	$HCOOH$	Insect stings
2	Acetic acid	Ethanoic acid	CH_3COOH	Vinegar
3	Propionic acid	Propanoic acid	CH_3CH_2COOH	Preservative for stored grains
4	Butyric acid	Butanoic acid	$CH_3(CH_2)_2COOH$	Butter
5	Valeric acid	Pentanoic acid	$CH_3(CH_2)_3COOH$	Valerian
6	Caproic acid	Hexanoic acid	$CH_3(CH_2)_4COOH$	Goat fat
7	Enanthic acid	Heptanoic acid	$CH_3(CH_2)_5COOH$	
8	Caprylic acid	Octanoic acid	$CH_3(CH_2)_6COOH$	Coconuts and breast milk
9	Pelargonic acid	Nonanoic acid	$CH_3(CH_2)_7COOH$	Pelargonium
10	Capric acid	Decanoic acid	$CH_3(CH_2)_8COOH$	Coconut and Palm kernel oil
11	Undecylic acid	Undecanoic acid	$CH_3(CH_2)_9COOH$	
12	Lauric acid	Dodecanoic acid	$CH_3(CH_2)_{10}COOH$	Coconut oil and hand wash soaps
13	Tridecylic acid	Tridecanoic acid	$CH_3(CH_2)_{11}COOH$	
14	Myristic acid	Tetradecanoic acid	$CH_3(CH_2)_{12}COOH$	Nutmeg
15	Pentadecanoic acid		$CH_3(CH_2)_{13}COOH$	
16	Palmitic acid	Hexadecanoic acid	$CH_3(CH_2)_{14}COOH$	Palm oil
17	Margaric acid	Heptadecanoic acid	$CH_3(CH_2)_{15}COOH$	
18	Stearic acid	Octadecanoic acid	$CH_3(CH_2)_{16}COOH$	Chocolate, waxes, soaps, and oils
19	Nonadecylic acid		$CH_3(CH_2)_{17}COOH$	Fats, vegetable oils, pheromone
20	Arachidic acid	Icosanoic acid	$CH_3(CH_2)_{18}COOH$	Peanut oil

Source: https://en.wikipedia.org/wiki/Carboxylic_acid.

Table 3. Properties of some acids, arranged in order of decreasing acid taste and with tartaric acid as reference

Acid	Properties of 0.05N solutions					
	Taste	Total acid g/L	pH	Ionisation constant	Taste sensation	Found in
Hydrochloric	+1.43	1.85	1.70	-	-	-
Tartaric	0	3.75	2.45	1.04×10^{-3}	Hard	Grape
Malic	-0.43	3.35	2.65	3.9×10^{-4}	Green	Apple, pear, prune, grape, cherry, apricot
Phosphoric	-1.14	1.65	2.25	7.52×10^{-3}	Intense	Orange, grapefruit
Acetic	-1.14	3.00	2.95	1.75×10^{-5}	Vinegar	-
Lactic	-1.14	4.50	2.60	1.26×10^{-4}	Sour, tart	-
Citric	-1.28	3.50	2.60	8.4×10^{-4}	Fresh	Berries, citrus, pineapple
Propionic	-1.85	3.70	2.90	1.34×10^{-5}	Sour, cheesy	-

Source: deMan (1999).

2.1.2. Dicarboxylic Acids

Dicarboxylic acids ($HO_2C\text{-}R\text{-}CO_2H$) is an organic compound which contains two carboxyl functional groups (-COOH). They have similar chemical properties to monocarboxylic acids. However, they have two dissociation constants; one for the dissociation into a mono anion and the second into a di anion (McMurry, 2008). The first ionization constant of dicarboxylic acids is usually larger than their monocarboxylic analogues because there are two potential sites for ionization making the effective concentration of the carboxyl group twice as large. Dicarboxylic acids are widely used in industries for the production of polymers (adipic acid), as food preservatives (oxalic acid) and as amino acids in human body (glutamic acid). Common examples of dicarboxylic acids and their dissociation constants are shown in Table 4 below (Nollet and Toldra, 2012; Huang et al., 2010).

Dicarboxylic acids can take various configurations depending on whether they are linear, branched chain, unsaturated or aromatic. Dicarboxylic acids are found naturally in plants and animals sources but in very little amount. Most dicarboxylic acids used industrially are synthetically produced from the oxidations of fatty acids.

Table 4. Example of dicarboxylic acids

Organic acid	Chemical formula	pKa_1	pKa_2
Oxalic acid	$C_2H_2O_4$	1.25	4.27
Maleic acid	$C_4H_4O_4$	1.91	6.33
Malonic acid	$C_3H_4O_4$	2.85	5.05
Fumaric acid	$C_4H_4O_4$	3.05	4.49
Malic acid	$C_4H_6O_5$	3.46	5.10
Ascorbic acid	$C_6H_8O_6$	4.10	11.79
Succinic acid	$C_4H_6O_4$	4.21	5.64
Adipic acid	$C_6H_{10}O_4$	4.41	5.41
Glutaric acid	$C_5H_8O_4$	4.34	5.41
Tataric acid	$C_4H_6O_6$	3.03	4.36

2.2. Citric Acid

The name citric acid was first derived from the latin word *citrus,* the citron tree that bears fruit which resembles a lemon. Citric acid was first extracted from lemon juice in 1784 by a Swedish chemist called Carl Scheele (Mattey and Kristiansen, 1999). Citric acid as shown in Figure 1, has been synthesized from glycerol and dichloroacetone by: (i) treating with hydrogen cyanide and hydrochloric acid to give dichloroacetonic acid, (ii) conversion of the resultant dichloroacetonic acid to dicyano-acetonic acid, (iii) with potassium cyanide which yields citric acid on hydrolysis (iv).

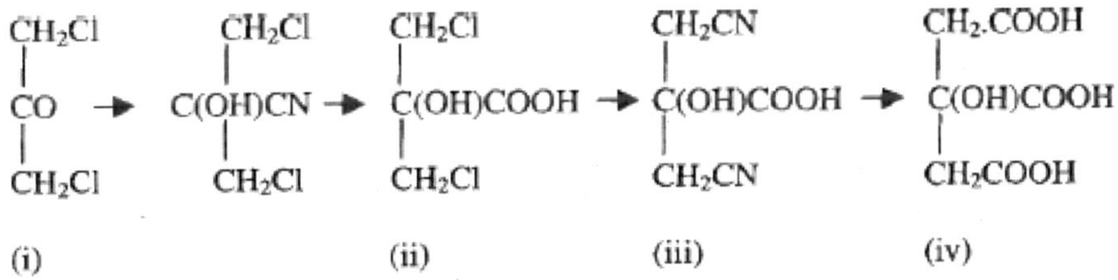

Figure 1. Synthesis of Citric acid. Source: Mattey and Kristiansen (1999).

Citric acid has also been found to be manufactured using microorganisms. Many microbes are implicated in the accumulation of citric acid and they include strains of *Aspergillus niger, A. awamori, A. fonsecaeus, A. luchensis, A phoenicus, A. wentii, A. saitoi, A. lanosius, A. flavus, Absidia* sp., *Acremonium* sp., *Aschochyta* sp., *Botrytis* sp., *Eupenicillium* sp., *Mucor piriformis,*

Penicillium janthinellum, P. restrictum, Talaromyces sp., *Trichoderma viride* and *Ustulina vulgaris* (Mattey and Kristiansen, 1999). In 1917, Currie discovered strains of *A. niger* that produced citric acid when cultured in media with low pH values, high sugar levels and mineral salts. Prior to this *A. niger* was known to produce oxalic acid; the key difference was the low pH which, as we now know, suppressed both the production of oxalic acid, which would be toxic, and gluconic acid, which has a significantly higher production rate from sugar than citric acid.

Another method for the production of citric acid is the Koji process which involves the use of *Aspergillus* species. The carbohydrate source, which is principally starch and cellulose, is sterilized by steaming and the resulting semi-solid paste which is composed of 70% of water, at a pH of about 5.5, is inoculated by spraying on spores of *A. niger*. Additions of ferrocyanide or copper may be made. The outcome of this method leads to low yield due to the difficulty in controlling trace metals and other process parameters. Sufficient cellulases and amylases are produced to break down the substrate, though the low yields may reflect the rate of limitation (Mattey and Kristiansen, 1999).

Citric acid has found application in food, beverages, pharmaceuticals and industrial fields with its application dependent on acidity, flavour and salt formation. Chemically citric acid is 2-hydroxy-1,2,3-propane tricarboxylic acid (CAS No. 77-92-9) composed of three p*K*a values at pH 3.1, 4.7 and 6.4. Citric acid forms a wide range of metallic salts including complexes with copper, iron, manganese, magnesium and calcium. Several studies point out the effectiveness of citric acid and its salts in various food systems. These salts are the reason for its use as a sequestering agent in industrial processes and as an anticoagulant blood preservative (Table 5). The metallic salts are also the basis of its antioxidant properties in fats and oils where it reduces metal-catalysed oxidation by chelating traces of metals such as iron. Some esters of citric acid of a range of alcohols have been identified and they include the triethyl, butyl and acetyltributyl esters used as plasticizers in plastic films. Monostytryl citrate is also used as an antioxidant in oils and fats (Mattey and Kristiansen, 1999).

Citric acid is a hydroxy tricarboxylic acid produced naturally by various plants. Citric acid is water soluble, approved for direct addition to multiple foods, is affirmed as generally regarded as safe (GRAS) and is approved for use in the manufacture of fresh and processed meats and poultry at concentrations specific to its purpose (USDA-FSIS, 2010). Acceptable daily intake for humans is presented in Table 6. Citric acid and its salts have demonstrated efficacy for pathogen control in both fresh and processed meat and poultry, but their usage is potentially limited by possible negative sensory impact and the need for low pH maintenance for optimum antimicrobial activity (Mani-Lopez et al., 2012). However, citric acid does not fit under the description of classic weak organic acids, that is, the lipophilic, undissociated acids. Therefore citric acid acts more as a chelator, exerting its antibacterial activity by sequestering metal ions such as Ca^{2+}, Mg^{2+}, and Fe^{3+} from the external medium required for bacterial homeostasis (Brul and Coote, 1999; Stratford and Eklund, 2003; Theron and Lues, 2011). Similar to lactic acid, citric acid can also act as a permeabilizing agent of the outer membrane of Gram-negative bacteria, as well as a potentiator of the effect of other antibacterial agents (Brul and Coote, 1999; Lianou and Koutsoumanis, 2012).

Table 5. Applications of citric acid

Industry	Property	Use
Food		
Beverages	Acidulant	Flavouring
Jellies, jams	Flavouring	Acidulant
Fats and oils	Antioxidant	Metal complexing
Frozen foods	Antioxidant	
Pharmaceutical		
Effervescent	Acid	Flavor
Vitamins	Antioxidant	
Anticoagulants	Sequestering	Buffering
Iron preparations	Salt formation	
Cosmetics	Buffering	Antioxidant
Industrial		
Cleaning (metals)	Sequestering	
Detergents	Buffering	Sequestering
Photographic	Buffering	
Primer binding	Sequestering	
Polymerization	Sequestering	

Source: Mattey and Kristiansen (1999).

Table 6. Acceptable daily intake for humans

Organic acids	Limitations (mg/kg body weight)	
	Unconditional	Conditional
Acetic	Not limited	
Acetate, Ca^{2+}, K^{+}, Na^{+}	Not limited	
Sodium diacetate	0 – 15	
Citric	Not limited	
Citrate, Ca^{2+}, K^{+}, Na^{+}	Not limited	
Lactic	Not limited	
DL-Lactic		0 – 100
Lactate, Ca^{2+}, K^{+}, NH^{4+}, Na^{+}	Not limited	
Malic	Not limited	
Propionic	Not limited	
Propionate, Ca^{2+}, K^{+}, Na^{+}	Not limited	
Tartaric	0 – 30	
Tartrate, K^{+}, Na^{+}	0 – 30	

Source: Mani-Lopez et al. (2012).

2.3. Acetic Acid

Acetic acid, as well as other OAs, has been known for ages. The Sumerians (2900 – 1800 BCE) used vinegar as a condiment, a preservative, an antibiotic and a detergent (Anon 2012). The acid is used in food preservation because of its effects on bacteria. Other carboxylic acids related to acetic acid are formic acid and propionic acid. The related compounds include: acetaldehyde, acetamide, acetic anhydride, acetonitrile, acetyl chloride, ethanol, ethyl acetate, potassium acetate, sodium acetate, and thioacetic acid (http://www.newworldencyclopedia.org/entry/Acetic_acid).

2.3.1. Chemical Properties and Thermochemistry of Acetic Acid

Acetic acid (ethanoic acid) CH_3COOH is a weak OA with other names as vinegar (diluted form), hydrogen acetate and methane carboxylic acid. It has a pKa of 4.7 - a logarithmic measure of the acid dissociation constant that categorizes the strength of an acid; the lower or more negative the number, the stronger and more dissociable the acid (http://www.newworldencyclopedia.org/entry/Acetic_acid). In terms of acidity (Figure 2), the hydrogen (H) atom in the carboxyl group (−COOH) in carboxylic acids can be given off as an H+

ion (proton), giving its acidic character. It is effectively a monoprotic acid in aqueous solution. Its conjugate base is acetate (CH_3COO-). Some other properties and thermochemistry of acetic acid are shown in Table 7.

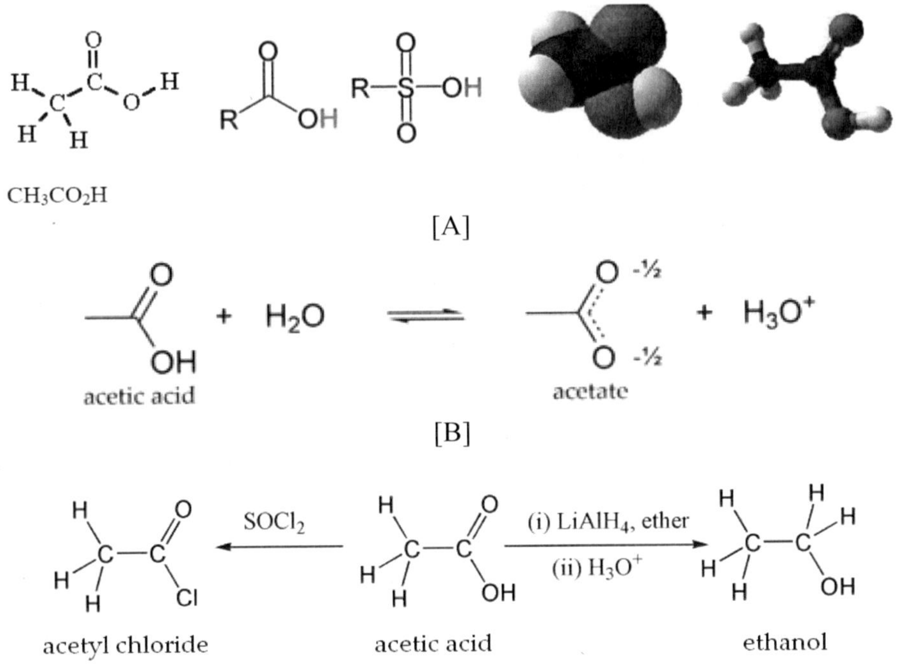

[C]

Source: http://www.chem.purdue.edu/gchelp/molecules/ch3co2h.html. [B] Acidity of acetic acid. Source: http://www.newworldencyclopedia.org/entry/Acetic_acid. [C] Two typical organic reactions of acetic acid. Source: https://en.wikipedia.org/wiki/Organic_acid.

Figure 2. [A] From left to right: General formula of Acetic acid, Carboxylic acid, Sulfonic acid, Spacefill model and Ball and stick model. The acidic hydrogen in each molecule is coloured red.

2.3.2. Mechanisms of Acetic Acid Antimicrobial Action

The mechanism of inactivation by weak OAs lays down in the ability of undissociated form of OA to penetrate through the cell membrane, and to dissociate inside the cell, resulting in decreased intracellular pH value, which is essential for the control of ATP synthesis, RNA and protein synthesis, DNA replication and cell growth (Yasothai and Giriprasad, 2015a). Beside the decrease in intracellular pH, the perturbation of the membrane functions by organic acid molecule might be also responsible for the microbial inactivation. The high concentration of anions (due to dissociation) inside the cells might result in an increased osmolarity and consequently to the metabolic perturbation (Hirshfield et al., 2003). As for other non-thermal inactivation treatments, the microbial sub-lethal injury might occur when the decontamination with organic acids is applied (Lee et al., 2002; Liao et al., 2003). Alexandrou et al. (1995) reported that weak OAs such as acetic and lactic acid showed greater ability to inflict the subpopulation of sub-lethal injured cells than stronger hydrochloric acid.

Table 7. Some properties and thermochemistry of acetic acid

Properties		Thermochemistry	
Chemical formula	$C_2H_4O_2$	Specific heat capacity (C)	123.1 J K^{-1} mol^{-1}
Molar mass	60.05 g·mol^{-1}	Std molar entropy (S^{o}_{298})	158.0 J K^{-1} mol^{-1}
Appearance	Colourless liquid	Std enthalpy of formation ($\Delta_f H^{o}_{298}$)	-483.88 - 483.16 kJ mol^{-1}
Odor	Pungent/Vinegar-like	Std enthalpy of combustion($\Delta_c H^{o}_{298}$)	-875.50 - 874.82 kJ mol
Density	1.049 g cm^{-3}		
Melting point	16 to 17°C; 61 to 62°F; 289 to 290 K		
Boiling point	118 to 119°C; 244 to 246°F; 391 to 392 K		
Solubility in water	Miscible		
log P	-0.322		
Acidity (pK_a)	4.76		
Basicity (pK_b)	9.24 (basicity of acetate ion)		
Refractive index(n_D)	1.371		
Viscosity	1.22 mPa s		
Dipole moment	1.74 D		

Source: https://en.wikipedia.org/wiki/Acetic_acid.

The key basic principle on the mode of action on bacteria is that non-dissociated (non-ionized) acid can penetrate the bacteria cell wall and disrupt the normal physiology of certain types of pH-sensitive bacteria, meaning that they cannot tolerate a wide internal and external pH gradient. Among those bacteria are *Escherichia coli*, *Salmonella spp.*, *Clostridium. perfringens*, *Listeria monocytogenes*, and *Campylobacter species*. Upon passive diffusion of OAs into the bacteria, where the pH is near or above neutrality, the acids will dissociate and lower the bacteria internal pH, leading to situations that will impair or stop the growth of bacteria. Thereafter, the anionic part of the OAs, which cannot escape the bacteria in its dissociated form, will accumulate within the bacteria and disrupt many metabolic functions, leading to osmotic pressure increase, incompatible with the survival of the bacteria (https://en.wikipedia.org/wiki/Organic_acid). The state of an acid (undissociated or dissociated) is extremely important in determining their capacity to inhibit the growth of bacteria.

A decrease in the pH is one of the factors affecting the action of OA solutions (Van Netten et al., 1994; Yasothai and Giriprasad, 2015a). Also the Gram-positive bacteria are more susceptible to the action of compounds interfering with the transport of ions across the cell (Raftari et al., 2009). The incorporation of acids into food can shorten sterilization times for heat treatment, owing to the lowered heat resistance of microorganisms in foods with increased acidity. The toxicology, antimicrobial properties, application, and regulatory status are different for each food acid. The carboxylic acids are more polar than lipophilic acids and are traditionally used in foods or their secondary effects rather than for their ability to inhibit microbial growth. The presence of acid can effectively inhibit germination and outgrowth of spores that survive the thermal process. Salt, sugar, and occurring agents in conjunction with acids (hurdle technology) serve to further decrease processing times. These interactions ensure food sterility, but the processing time would also aid in preserving the palatability of the product (Doores, 2005).

2.3.3. Organic Salts as Preservatives

Some organic salts are used as preservatives in food products and an example is potassium acetate. They prevent spoilage by inhibiting the growth of bacteria and fungi. Calcium and sodium propionate, for example, are added to processed cheese and bakery goods; sodium benzoate is added to cider, jellies, pickles, and syrups; and sodium sorbate and potassium sorbate are added to fruit juices, sauerkraut, soft drinks, and wine (Anon, 2012). There preservatives are found on ingredient labels at groceries shop. OAs are also used as dietary acidifiers for pigs or poultry (Dibner and Buttin, 2002) and have several additional effects that go beyond those of antibiotics; including reduction in digesta pH, increased pancreatic secretion, and trophic effects on the gastrointestinal mucosa.

2.3.4. Production of Acetic Acid

Prehistoric people likely made acetic acid when their fermentation reactions went awry and produced vinegar instead of wine (Anon, 2012). Acetic acid is produced both synthetically and by bacterial fermentation. The biological route accounts for only about 10% of world production, but it remains important for vinegar production, as many of the world food purity laws stipulate that vinegar used in foods must be of biological origin. About 75% of acetic acid made for use in the chemical industry is made by methanol carbonylation. Alternative methods account for the rest (Yoneda, 2001). Total worldwide production of virgin acetic acid is estimated at 5 Mt/a (million metric tons per year), approximately half of which is produced in the United States. European production stands at approximately 1 Mt/a which is presently declining, with 0.7 Mt/a produced in Japan. Another 1.5 Mt are recycled each year, bringing the total world market to 6.5 Mt/a (C&EN, 2005; Malveda, 2007). Figure 3 shows the world consumption of acetic acid in 2014.

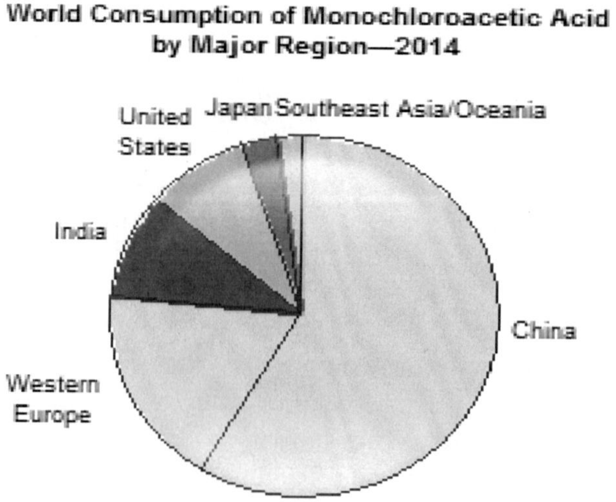

Source: CEH (2013).

Figure 3. Pie chart showing world consumption of acetic acid.

Several organic or inorganic salts are produced from acetic acid, including: Sodium acetate - used in the textile industry and as a food preservative (E262); Copper(II) acetate - used as a pigment and a fungicide; Aluminium acetate and iron(II) acetate - used as mordants for dyes; and Palladium(II) acetate - used as a catalyst for organic coupling reactions such as the Heck reaction (http://www.newworldencyclopedia.org/entry/Acetic_acid).

2.3.5. Oxidative Fermentation in Acetic Acid Production

Acetic acid, in the form of vinegar, has been made by bacteria of the genus *Acetobacter.* These bacteria, with sufficient oxygen, can produce vinegar from a variety of alcoholic foodstuffs. Commonly used feeds include apple cider, wine, and fermented grain, malt, rice, or potato mashes (http://www.newworldencyclopedia.org/entry/Acetic_acid). The overall chemical reaction facilitated by these bacteria is shown in Equation 1:

$$C_2H_5OH + O_2 \rightarrow CH_3COOH + H_2O \quad (1)$$

A dilute alcohol solution inoculated with *Acetobacter* and kept in a warm, airy place will become vinegar over the course of a few months. Industrial vinegar-making methods accelerate this process by improving the supply of oxygen to the bacteria. The first batches of vinegar produced by fermentation probably followed errors in the winemaking process. If it is fermented at too high a temperature, acetobacter will overwhelm the yeast naturally occurring on the grapes. As the demand for vinegar for culinary, medical, and sanitary purposes increased, vintners quickly learned to use other organic materials to produce vinegar in the hot summer months before the grapes were ripe and ready for processing into wine. However, this method was slow (http://www.newworldencyclopedia.org/entry/Acetic_acid).

One of the first modern commercial processes was the "fast method" or "German method," first practiced in Germany in 1823. In this process, fermentation takes place in a tower packed with wood shavings or charcoal. The alcohol-containing feed is trickled into the top of the tower, and fresh air supplied from the bottom by either natural or forced convection. The improved air supply in this process cut the time to prepare vinegar from months to weeks.

Vinegar is now made in submerged tank culture, first described in 1949 by Otto Hromatka and Heinrich Ebner. In this method, alcohol is fermented to vinegar in a continuously stirred tank, and oxygen is supplied by bubbling air through the solution. Using this method, vinegar of 15 percent acetic acid can be produced in only 2 – 3 days.

2.3.6. Anaerobic Fermentation in Acetic Acid Production

Some species of anaerobic bacteria, including several members of the genus *Clostridium,* can convert sugars to acetic acid directly, without using ethanol as an intermediate. The overall chemical reaction by these bacteria may be represented in Equation 2 as:

$$C_6H_{12}O_6 \rightarrow 3CH_3COOH \qquad (2)$$

More interestingly from the point of view of an industrial chemist, many of these acetogenic bacteria can produce acetic acid from one-carbon compounds, including methanol, or a mixture of carbon dioxide and hydrogen (Equation 3):

$$2CO_2 + 4H_2 \rightarrow CH_3COOH + 2H_2O \qquad (3)$$

This ability of *Clostridium* to utilise sugars directly, or to produce acetic acid from less costly inputs, means that these bacteria could potentially produce acetic acid more efficiently than ethanol-oxidisers like *Acetobacter.* However, *Clostridium* bacteria are less acid-tolerant than *Acetobacter.* Even the most acid-tolerant *Clostridium* strains can produce vinegar of only a few per cent acetic acid, compared to some *Acetobacter* strains that can produce vinegar of up to 20% acetic acid. It is more cost-effective to produce vinegar using *Acetobacter* than to produce it using *Clostridium* and then concentrating it (http://www.newworldencyclopedia.org/entry/Acetic_acid).

2.3.7. Applications of Acetic Acid

Acetic acid is a chemical reagent for the production of many chemical compounds. The largest single use of acetic acid is in the production of vinyl acetate monomer (VAM), closely followed by acetic anhydride and ester production. (http://www.newworldencyclopedia.org/entry/Acetic_acid). The major use of acetic acid is for the production of VAM. This application consumes approximately 40 - 45% of the world's production of acetic acid. The reaction (Equation 4) is of ethylene and acetic acid with oxygen over a palladium catalyst.

$2H_3C\text{-}COOH + 2C_2H_4 + O_2n \rightarrow 2H_3C\text{-}CO\text{-}O\text{-}CH{=}CH_2 + 2H_2O$ **4**

acetic acid ethylene vinyl acetate

Vinyl acetate can be polymerised to polyvinyl acetate or to other polymers, which are applied in paints and adhesives. The condensation product of two molecules of acetic acid is acetic anhydride. The worldwide production of acetic anhydride is a major application, and uses approximately 25 - 30% of the global production of acetic acid. Acetic anhydride may be produced directly by methanol carbonylation by passing the acid.

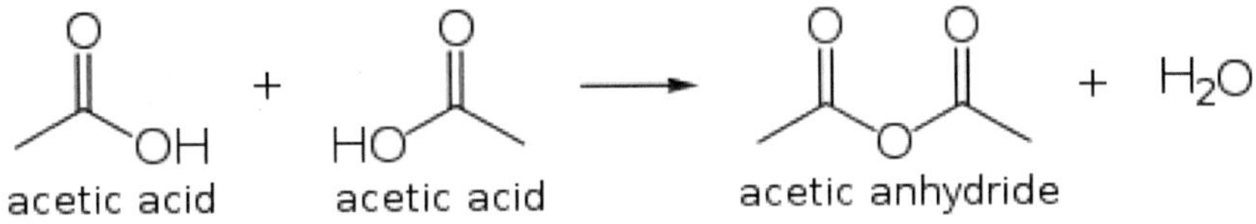

Figure 4. Condensation of acetic acid.

Acetic anhydride is a strong acetylation agent. As such, its major application is for cellulose acetate, a synthetic textile used for photographic film. Acetic anhydride is also a reagent for the production of aspirin, heroin, and other compounds.

2.3.8. Acetic Acid Use as Solvent

Acetic acid is often used as a solvent for reactions involving carbocations, especially in analytical chemistry. In other applications, dilute solutions of acetic acids are also used for their mild acidity. Examples in the household environment include the use in a stop bath during the development of photographic films, and in descaling agents to remove limescale from taps and kettles. Equivalently, acetic acid is used as a spray-on preservative for livestock silage, to discourage bacterial and fungal growth (http://www.newworldencyclopedia.org/entry/Acetic_acid).

2.3.8.1. Vinegar

Vinegar, typically 4 - 18% (by mass) acetic acid, is used directly as a condiment, and in the pickling of vegetables and other foods. Table vinegar (Figure 5) tends to be more dilute (4 – 8% acetic acid solution), while commercial food pickling generally employs more concentrated solutions (http://www.newworldencyclopedia.org/entry/Acetic_acid). Vinegar is the oldest and most well-known application of acetic acid (https://en.wikipedia.org

/wiki/Acetic_acid). The volume of acetic acid used in vinegar is comparatively small (http://www.newworldencyclopedia.org/ entry/Acetic_acid).

Commercial vinegar is produced either by fast or slow fermentation processes. The slow methods are used with traditional vinegars, and fermentation proceeds slowly over the course of months or a year. The longer fermentation period allows for the accumulation of a nontoxic slime composed of acetic acid bacteria. Fast methods add vinegar-bacterial culture to the source liquid before oxygenation promoting fast fermentation. In fast production processes, vinegar may be produced in 20 h to 3 days. https://en.wikipedia.org/ wiki/Vinegar.

Source: https://en.wikipedia.org/wiki/Vinegar.

Figure 5. Table vinegar.

Table 8. Content of OAs in vinegars (g/L) based on the heights of the peaks (Aguiar et al., 2005)

Amples of vinegars	Acid							
	Malic	Acetic	Propionic	Malonic	Lactic	Tartaric	Citric	Succinic
White wine and alcohol vinegar A	0.20 ± 0.02	56.2 ± 0.8	nd	0.30 ± 0.01	nq	0.24 ± 0.04	0.12 ± 0.02	nq
Red wine and alcohol vinegar A	nd	49 ± 2	nq	0.41 ± 0.02	0.20 ± 0.03	0.218 ± 0.006	0.127 ± 0.002	nq
White wine (10%) and alcohol (90%) vinegar B	nd	54.2 ± 0.8	nq	nq	1.62 ± 0.06	nq	nq	0.191 ± 0.004
White wine and alcohol vinegar C	nq	52 ± 1	nq	nq	0.19 ± 0.01	0.134 ± 0.003	nq	nq
Apple vinegar C	nd	50 ± 2	0.54 ± 0.01	0.17 ± 0.05	2.2 ± 0.2	nq	0.55 ± 0.04	1.07 ± 0.08
Apple vinegar D	0.150 ± 0.009	50.6 ± 0.8	0.319 ± 0.004	nd	0.63 ± 0.03	nd	0.163 ± 0.002	0.74 ± 0.02
Red wine and alcohol vinegar C	nq	44 ± 1	nq	nq	0.150 ± 0.005	0.126 ± 0.003	nq	nq
Rice vinegar E	nd	52 ± 2	0.28 ± 0.07	nq	0.45 ± 0.01	nd	nq	0.217 ± 0.006
Alcohol vinegar C	nd	52.0 ± 0.8	nq	nd	nq	nq	nq	nd

nd - not detected; nq - identified, but not quantified.

Some OAs found in vinegars can be originally from the grape or produced in the alcoholic, acetic or malolactic fermentations. The total acidity is expressed in acetic acid, the major organic acid in vinegars. Tartaric acid is major acid in wines and vinegars of grape wines, because it is original from the own fruit. Aguiar et al. (2005) showed that in all the vinegars samples analysed (Table 8) acetic acid was the most abundant component (40 - 56 g/L) followed by lactic acid (0.2 - 2.2 g/L). They detected tartaric acid in all the wine vinegars. OAs are one of the major phytochemicals in vegetables and responsible for food taste and odor. Different OAs are analysed in fruits and cereals, but least analysed in vegetables and spices (Priecina and Karklina, 2015). OAs are typical products of cell metabolism and all occur naturally in a variety of vegetables and animal substrates, and can be present either as constituents of foods as a result of normal biochemical metabolic processes, direct addition as acidulates, hydrolysis or bacterial growth, or can be later added directly or indirectly in products (Theron and Rykers, 2010).

They are important to biological processes, since they are involved in various fundamental pathways in plant and animal metabolism and catabolism as intermediate or final products (Gonzalez and Gonzalez, 2012). Most raw vegetables are unpalatable and can undergo growing quality changes associated with enzymatic reactions; therefore, it is necessary to process vegetables to increase their shelf life and improve their eating quality. During processing, vegetables are often subjected to mechanical (peeling, cutting, mixing, homogenisation, coring, etc.) and thermal (blanching) treatments. These treatments and the sequences of performing them could influence the stability of vitamin C and other important nutrients during the treatments themselves or during subsequent processing, straight to the occurrence of chemical and enzymatic oxidation reactions (Munyaka et al., 2010).

2.4. Benzoic Acid

Benzoic acid (C_6H_5COOH) is one of the oldest and most commonly used food preservative (Barbosa-Canovas et al., 2003). It is widely used as a preservative for food and beverages. Benzoic acid occur naturally at a high level in many fruits such as cranberries, plums, cinnamon and prunes. Indeed, some berries, such as cloudberries, contain so much benzoic acid that they can be stored for long periods without bacterial or fungal spoilage (Piper, 2011; Huang et al., 2012). Its preservative effectiveness depends on the acidity of the food (Clayden et al., 2012). It has been widely used as preservatives in soft drinks, fruit drink and pickles. Common levels of benzoic acid found in different food substances is presented in Table 9.

Table 9. Levels of benzoic acid in selected food materials

Food Items	Levels (mg/kg)	Reference
Milk	Traces – 6	Sieber et al. (1989)
Cheese	12 – 40	Sieber et al. (1989)
Fruits (excluding *Vaccinium* species)	Traces – 40	Sieber et al. (1989)
Potatoes, beans and cereals	Traces – 14	Sieber et al. (1989)
Soya flour and nuts	1.2-11	Sieber et al. (1989)
Honey	10-100	Steeg and Montang, 1987

Benzoic acid, a white crystalline solid is slightly soluble in water and this restricts its use as a preservative in food industries. Sodium benzoate a salt of derivative of the acid (C_6H_5COONa) is more soluble in water than the benzoic acid and it is commonly used as a preservative in food industries than benzoic acid because of its greater solubility in aqueous solution. Benzoates are derived from a neutralization reaction with benzoic acid and are more commonly used as food preservatives than the acid. Benzoic acid was first extracted from the resin of trees belonging to the Styrax genus known as gum benzoin (Crampton, 2016; Pastrorova et al., 1997). Figure 6 shows the conversion of benzoic acid to sodium benzoate.

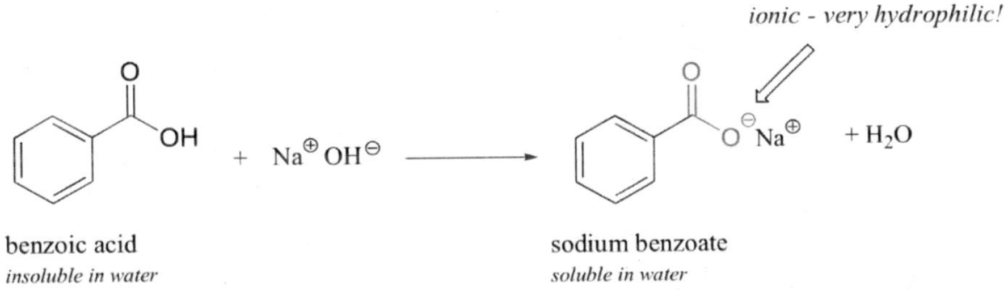

Figure 6. Conversion of benzoic acid to sodium benzoate.

Benzoic acid is produced industrially by the partial oxidation of toluene with oxygen as shown in the Figure 7 below.

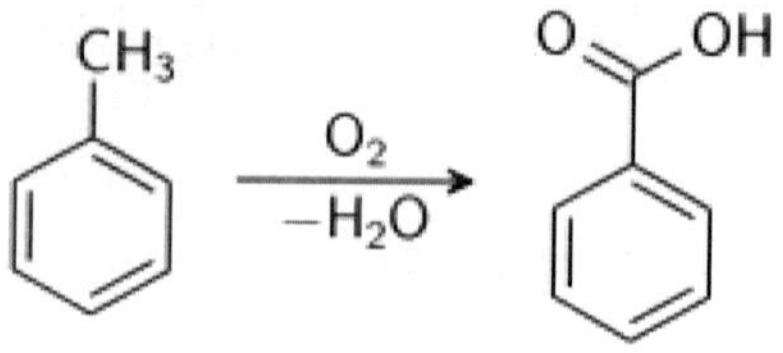

Figure 7. Benzoic acid production.

Sodium benzoate was the first chemical preservative allowed by the FDA for food products. It is readily converted to benzoic acid in acidic medium. It has been successfully applied as a preservative for foods and beverages in a range of pH < 4.5. The undissociated acid has antimicrobial activity and this makes it well suited for use in acid foods (Bilau et al. 2008). Apart from it acting as a bactericide it also act as an antifungal agent in fruit juices (Barbosa-Canovas et al., 2003). Benzoic acid has also found use as a preservative in toothpastes, mouthwashes, ointment, cosmetics, shampoos and pharmaceutical industries for the prevention of microbial growth (Otero-Losada, 2003; Crampton, 2016). It prevents microorganisms by preventing glucose from fermenting (Crampton, 2016).

Suhr and Nielsen (2004) indicated that sodium benzoate is mainly applied as a preservative of fruit and fruit juices. Poulter (2007) highlighted that sodium benzoate have been used as preservatives in soft drinks, baked goods and lollipops. Barbosa-Canovas et al. (2003) reported that benzoic acid derivatives and parabenzoates are used in fruit juices, chocolate syrup, pickled vegetables, pie fillings and cheese. Warth (1991) indicated that benzoic acid inhibits the growth of mold, yeast and some bacteria. Benzoic acid are usually added to a food substrate either directly or indirectly. The mechanism starts with the absorption of benzoic acid into the cell. Acidic food and beverage like fruit juice (citric acid), sparkling drinks (carbon dioxide), soft drinks (phosphoric acid), pickles (vinegar) or other acidified food are preserved with benzoic acid and its derivatives. The levels of benzoic acids with its derivatives in food is usually in the range of 0.05 - 0.1% (WHO, 2000).

Benzoic acid and its salts (usually called benzoates) are used as food additives which is used for the preservation of various food items from various pathogenic organisms including bacteria, yeasts and fungi (Bilau et al. 2008). Although, yeasts are inhibited by benzoate to a greater extent than moulds and bacteria (FAO, 1995). Most yeasts and molds are inhibited by 0.002–0.07% benzoic acid (Luck, 1977). Benzoic acid occurs naturally in some plants and animals. They are also present in varying amount in various food items such as honey, milks and fruits (Sibier et al., 1995). The sodium, potassium and calcium salt of benzoic acid usually have higher antimicrobial effects when used on food substrate than benzoic acid.

2.4.1. Health Concerns About the Use of Benzoic Acid and Benzoates as Food Preservatives

The presence of benzoic acid and benzoates in relatively low amount does not constitute any health risks to the consumer, however higher levels of benzoic acid have been implicated for several health risks (Crampton, 2012). The consumption of foodstuffs that contains benzoic acid either naturally or added as a food preservative have been implicated as the major route of human exposure to benzoic acid (WHO, 2000). The benzoic acid derived from eating natural food substances is often in low concentrations and is not dangerous to health. Irritation of eyes, skin and lungs have been linked with the use of benzoates as preservatives in food items (Sibier et al., 1995; WHO, 2000). The use of benzoic acid with ascorbic acid in soft drinks has led to the formation of benzene which is a carcinogen. Some soft drinks have been reported to contain high levels of benzene above the recommended guideline value by Food and Drug regulation agency. Tfouni et al. (2002) highlighted several adverse effects in humans and animals associated with high dose of benzoic acid used as a food preservatives; some of such effects includes metabolic acidosis, convulsions and hyperpnoea. Asthma, rhinitids, urticarial in humans have also been linked with the use of sodium benzoates (Hannuksela and Haahtela, 1987; Juhlin, 1981; WHO, 2000).

2.4.2. Mode of Action of Benzoic Acid as Food Preservative

Generally the mechanisms of benzoic acid inhibition of growth in foodstuffs have been linked to its ability to reduce cytoplasmic pH in acidic condition which inhibits phosphofructokinase. Although some notorious species of yeast e.g., *Zygosaccharomyces bailii* (which deteriorates several acidic food) are known to inhibit the performance of benzoic acid even at high levels. This inhibition by yeast is by keeping the intracellular concentration of preservative anions low. Warth (1991) reported that for the inhibition of pH reduction also does not appear to be a prime cause of inhibition in *Z. bailii*, rather, the pattern of glycolytic intermediate changes suggests that ATP limitation was important. Several other mechanisms for the action of benzoic acid and other OAs for food preservation include: inhibition of essential metabolic rates, membrane disruptions, stress on intracellular pH homeostasis, and accumulation of toxic anions (Stratford and Anslow, 1998; Brul and

Coote, 1999; Holyoak et al., 1996; Eklund, 1985; Bracey et al., 1998; Salmond et al., 1984; Krebs et al., 1983). Figure 8 shows a typical resistance pattern of yeast cell to weak OAs as reported by Brul and Coote, (1999).

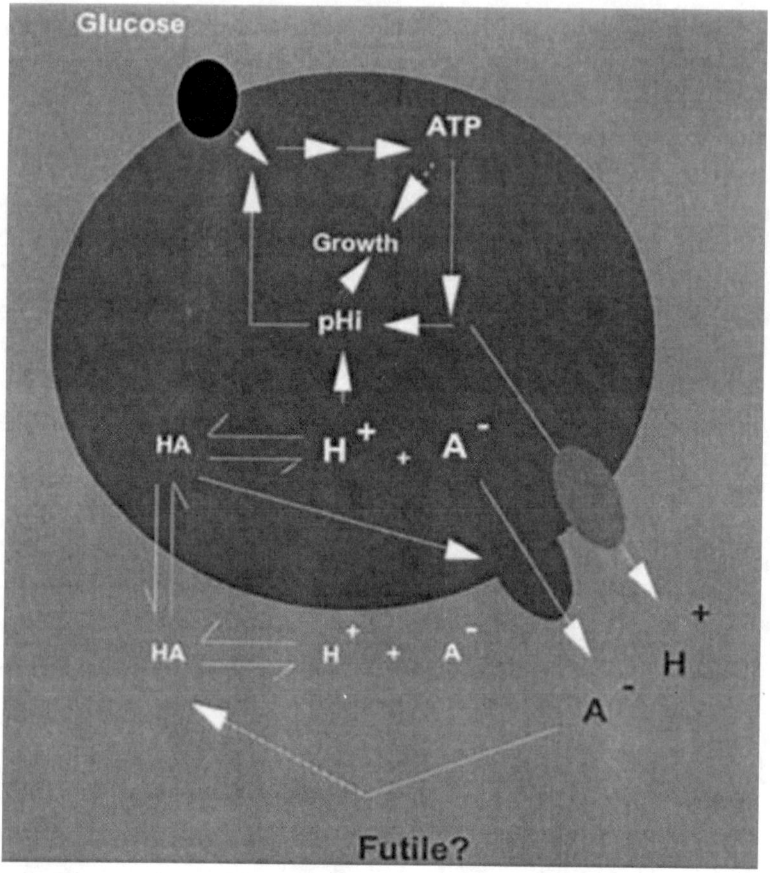

Figure 8. A schematic diagram of the stress response of a yeast cell challenged with weak organic acids (Piper et al., 1998). Shown are, a glucose transporter, the membrane located Pdr12 multidrug resistance pump active against anions of acetic, sorbic and benzoic acid, and the 1 plasma membrane P-type H -ATPase.

2.5. Sorbic Acid

Sorbic acid (2,4-hexadienoic acid) is an unsaturated monocarboxylic acid straight chain fatty acid with conjugated double bonds (Figure 9). Sorbic acid and its soluble salts (potassium and calcium sorbates) are widely used in food and pharmaceutical industries. Sorbic acid is known to effectively inhibit the activity of mold, yeast, aerobic bacteria and prevent the growth and reproduction of botulism, staphylococcus, salmonella and other harmful microbials (Ferrand et al., 2000; Fang et al., 2015; Saraiva et al., 2015).

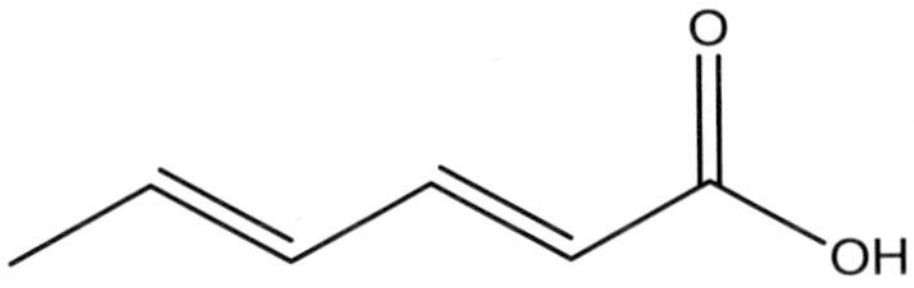

Figure 9. Sorbic acid.

The most acceptable mechanism for sorbic acid antimicrobial activity is the inactivation of enzymes necessary to sustain the vital activities of microorganisms by formation of covalent bonds between the double bonds of sorbic acid and thiol group (-SH) of enzymes, leading to cessation of metabolic activities in the microbial cells (Alagoz et al., 2015). However, some spoilage moulds like *Aspergillius niger* are able to decarboxylate sorbic acid to the volatile and less toxic 1,3-pentadiene with kerosene like odur. The resistance to sorbic acid by *A. niger* also permits growth of other sorbic acid-sensitive microbes (Straford et al., 2012).

Sorbic acid has low toxicity due to rapid metabolism by pathways similar to those of other fatty acids (Santini et al., 2009). Sorbic acid and its salts have an acceptable daily intake value of 25 mg/Kg body weight by Joint US Food and Agriculture Organization/World Health Organization expert committee on food additives (Ohtsuki et al., 2016). The low toxicity could be responsible for the high daily intake limit. Sorbic acid salts are preferred by the food industry over other preservatives like propionic acid and benzoic acid due to their inhibition of microbial growth at high pH (6.0 - 6.5); high acceptable intake of 25 mg/Kg body weight (benzoates have intake of 5 mg/Kg body weight); and little effect on human health.

3. Mode of Antimicrobial Action of OAs

Most OAs are capable of freely moving throughout bacterial cells, due to their simple structure and small molecular size (Theron and Lues, 2007). Thus, OAs are able to exhibit bacteriostatic or bactericidal properties depending on the physiological status of the target organism and the physicochemical characteristics of the external environment (Ricke, 2003). The antimicrobial action of weak OAs depends on factors such as the pH lowering effect; the extent of dissociation of the acid; and a specific effect associated with the acid molecule (Lianou and Koutsoumanis, 2012). Bacterial inactivation by weak OAs has been traditionally attributed to the ability of the lipophilic, undissociated acid molecules to penetrate the cytoplasmic membrane and dissociate inside the bacterial cell (Lianou and Koutsoumanis, 2012). Undissociated OAs possess high antimicrobial activity, much stronger than that exerted by their dissociated forms. The extent of dissociation which is the concentration of undissociated acid and its antimicrobial effectiveness in solution is determined by its pKa and the pH of the external medium (Birk et al., 2010; Helander et al., 1997; Ricke, 2003; Stratford and Eklund, 2003).

The undissociated state of the compounds are favoured by low pH values with the uncharged compounds crossing the cell membrane and gaining access to the cell (Birk et al., 2010; Brul and Coote, 1999). Upon entering the bacterial cell and encountering higher pH environment of the cytoplasm, the acid molecule dissociates and charged ions are released and accumulate inside the cell. The release and accumulation of charged ions inside the cell continues until equilibrium in accordance with the pH gradient across the membrane is reached thus resulting in reduced intracellular pH and disruption of the membrane proton-motive force (Brul and Coote, 1999; Lianou and Koutsoumanis, 2012). Therefore, factors responsible for the antimicrobial activity of OAs include membrane disruption, stress on intracellular pH homeostasis resulting in energy depletion and inhibition of essential metabolic reactions (Brul and Coote, 1999; Ricke, 2003; Lianou et al., 2012).

In solution, weak acid preservatives exist in a pH-dependent equilibrium between the undissociated and dissociated state. Preservatives have optimal inhibitory activity at low pH because this favours the uncharged, undissociated state of the molecule which is freely permeable across the plasma membrane and thus able to enter the cell. Therefore, the inhibitory action is classically believed to be due to the compound crossing the plasma membrane in the undissociated state. Subsequently, upon encountering the higher pH inside the cell, the molecule will dissociate resulting in the release of charged anions and protons which cannot cross the plasma membrane. Thus, the preservative molecule diffuses into the cell until equilibrium is reached in accordance with the pH gradient across the membrane resulting in the accumulation of anions and protons inside the cell (Brul and Coote, 1999). Therefore, inhibition of growth by weak acid preservatives has been proposed to be due to a number of actions including membrane disruption, inhibition of essential metabolic reactions, stress on intracellular pH homeostasis and the accumulation of toxic anions (Brul and Coote, 1999).

In microorganisms, as shown in Figure 10, the undissociated form of OA (HA) is diffusing through the microbial membrane when the pH of the cellular cytoplasm is higher than that of the surrounding environment. In order to maintain the internal pH (at neutral pH), active transport to efflux protons (H^+) is required. To export excess proton out of the cytoplasm, adenosine triphosphate is consumed at the expense of cellular activities, thereby depleting cellular energy after a period of time. Furthermore, the release of anions increases the osmotic pressure in the cytosol thereby deleterious to cytosolic enzymes. Therefore, continuous increase in cellular acidity damage or modifies the functionality of enzymes, structural proteins, and DNA. Few OAs (malic and citric acids) have been shown to efficiently destabilize the outer membrane by chelation or intercalation (Ter Beek et al., 2015; Back et al., 2009; Romano et al., 2015; Dishissha et al., 2015).

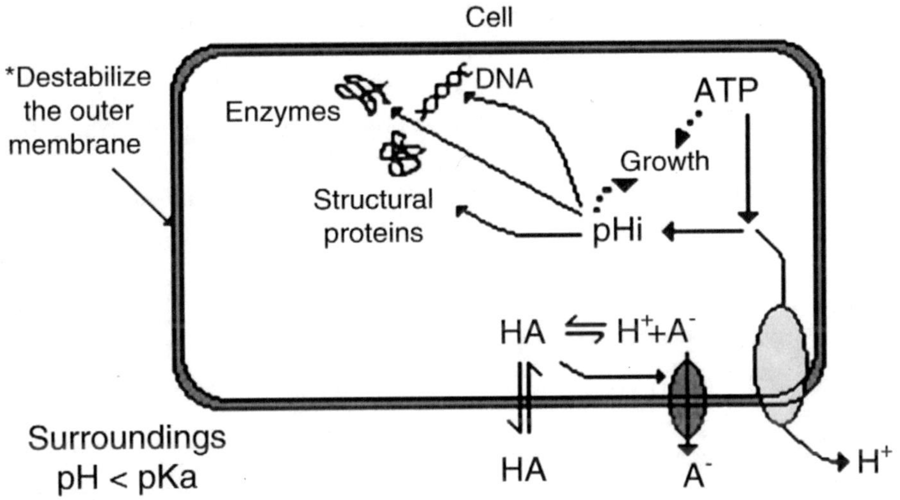

Figure 10. Mechanism of organic acid on microbial cells. Source: Mani-Lopez et al. (2012).

Lactic, acetic, citric, tartaric, fumaric, levulinic and peroxyacetic acids are said to have recognized potential as disinfectants in fruits and vegetables, especially those intended for preparation of RTE salads (Sapers, 2001; Gil et al., 2009; Ölmez and Kretzschmar, 2009; Raybaudi-Massilia et al., 2009b). Their industrial use as decontaminants of fresh produce is permitted provided they function as processing aids (Gil et al., 2009; Koutsoumanis and

Skandamis, 2013). Some of them, alone or in combination with other disinfectants, have been included in the formulation of commercial Sanitizers. However, their application are to be done under conditions that maximize effectiveness in microbial reduction and without adverse effects on sensory or nutrient quality.

The decontamination efficacy of OAs on both meat, vegetables and fruits surfaces depends on the application conditions, rinsing after sanitation, the concentration of acid, the characteristics of the food surface, strength of bacterial attachment and the internalization of cells in the food (Ölmez and Kretzschmar, 2009; Raybaudi-Massilia et al., 2009b; Skandamis et al., 2010; Koutsoumanis and Skandamis, 2013). Consequently, the antimicrobial activity of OAs is enhanced as the pH of the food is lowered to that of, or below, the pKa of the acid. The pKa here is defined as the acid dissociation constant. Reduction in pH results in a greater concentration of protonated acid, decreasing the polarity of the molecule and increasing diffusion of acid across the membrane and into the cytoplasm. However, the substitution of the free proton with a monovalent (Na^+, K^+) or multivalent (Ca^{2+}) cation significantly increases the solubility of OA in aqueous systems. Thus, a balance must be made between the need to maintain acid solubility with the need to achieve maximal activity via pH reduction (Mani-Lopez et al., 2012). Most recently, research has demonstrated that the interplay of all these mechanisms likely drives inhibition of microbes by OAs (Koczon, 2009; Mani-Lopez et al., 2012).

4. Future Considerations for Organic Acid Use in Food Processing and Preservation

OAs have wide range of potentially useful antimicrobial activity hence their use in food preservation. Concern has been raised on safe and effective techniques for applying OAs to prevention of bacterial growth in food. Theron and Lues, (2010) stated that although OAs have been used to counteract pathogens in food for many years, there is a glaring need to assess and improve their continued effectiveness and sustainability. They are of the view that there is also a growing demand for foods that are produced using milder treatments (less heat, salt, sugar, and chemicals), guides in the selection of appropriate OA for specific food products (Davidson et al., 2005) and newer technologies to prevent the growth of dangerous bacteria.

Complex issues regarding food preservation and safety have emerged (Davidson et al., 2005) and continue to emerge. With the increase in acid-tolerant and resistant microorganisms, increasing outbreaks of *Escherichia coli* O157:H7 and *Listeria monocytogenes,* strategies are needed to stem this trend or search for alternatives. It is also known that it is difficult to compare the activity of different acids because it is influenced by the physical chemistry of the microbial species, the growth conditions, and the phase of growth (Cherrington, 1991). This and other concerns have led to increasing research at industries and laboratories around the globe. The result of this is specific application regimen and other emerging strategies involving food antimicrobials of natural and synthetic origin, together with additional mechanisms of action involved in a number of naturally occurring antimicrobials.

Food industries face challenges to satisfy modern consumer trends, food legislation, and consumer and regulatory demands for improved food safety (Davidson et al., 2005). It is envisaged that scientists will continue to explore the continuous effectiveness of OAs as a natural preservative in most foodstuff as well as possible solutions from interdisciplinary nature of food science and technology that embraces microbiology, technology, biochemistry, biotechnology, and engineering. For sustainability and food security, search will continue on advances on lysozyme, naturally occurring antimicrobials from both animal and plant sources, hurdle technology (HT) approaches and mechanisms of action, resistance, and stress adaptation.

According to Anon (2016), HT is a method of ensuring that pathogens in food products can be eliminated or controlled by combining more than one approach. This means the food products will be safe for consumption, and their shelf life will be extended. These approaches can be thought of as "hurdles" the pathogen has to overcome if it is to remain active in the food. The right combination of hurdles can ensure all pathogens are eliminated or rendered harmless in the final product (Alasalvar, 2010; Lee, 2004). Leistner (2000) defined HT as an intelligent combination of hurdles which secures the microbial safety and stability as well as the organoleptic and nutritional quality and the economic viability of food products. The organoleptic quality of the food refers to their sensory property; that is, its look, taste, smell and texture. Examples of hurdles in a food system are high temperature during processing, low temperature during storage, increasing the acidity, lowering the water activity or redox potential, or the presence of preservatives. According to the type of pathogens and how risky they are, the intensity of the hurdles can be adjusted individually to meet consumer preferences in an economical way, without compromising the safety of the product (Alasalvar, 2010). Each hurdle aims to eliminate, inactivate or at least inhibit unwanted microorganisms. Common salt or OAs can be used as hurdles to control microbials in food. Many natural antimicrobials such as nisin, natamycin and other bacteriocins and essential oils derived from rosemary or thyme, also work effectively as hurdles (Anon, 2016).

There is a growing demand for foods that are produced using milder treatments and newer technologies to prevent the growth of dangerous bacteria (Theron and Lues, 2010). Novel applications for OAs include the following: emerging challenges, consumer satisfaction, optimizing OA application in animal feed, preservative combinations, antimicrobial packaging, optimizing commercial trials, new possibilities in minimally processed foods, alternatives to washing techniques, alternative application regimes, and recognizing the need in RTE foods. Equally, other novel compounds have now been approved for use like lysozyme, lactoferrin, ozone and several others.

The emergence of non-thermal food-preservation technologies to preserve fruit and vegetable juices are innovative and potentially useful alternatives to replace the use of chemical additives and intense heat treatments (Leite et al., 2016). The authors observed that such technologies as the incorporation of essential oils or their individual constituents into fruit and vegetable juices can effectively reduce or inhibit pathogenic and spoilage microorganisms. Although ultrahigh pressure (UHP), also known as pascalization, high hydrostatic pressure processing or ultra-high pressure processing (Yasothai and Giriprasad, 2015b), offer interesting possibilities for food processing and preservation, Smelt (1998) observed no specific effect of OAs apart from pH effects. He attributed this to the fact that pressure favours ionization and that OAs are particularly inhibitory in the undissociated form. It is however conceivable that under pressure the undissociated part might be more active. The pressures between 300 and 600 MPa have been found to inactivate yeasts, moulds and most vegetative bacteria including most infectious food-borne pathogens.

CONCLUSION

Despite the risk involved in the development of enhanced microbial resistance during use and application of OAs, when used and applied properly and at appropriate concentrations, chemical decontamination treatments are expected to constitute valuable pathogen control interventions. This greatly reduces the food safety risks associated with stress-adaptation phenomena in food produce. With the availability of many different chemical agents that can be utilized in decontamination systems, rotation of the use of differing agents over time within a certain food processing facility has therefore been proposed as a means of preventing selection for bacterial resistance (Samelis et al., 2001b; Lianou and Koutsoumanis 2012).

References

Aguiar, A., Nascimento, R. A., Ferretti, L. P. and Goncalves, A. R. (2005). Determination of organic acids and ethanol in commercial vinegar. *Braz. J. Food Technol., 5,* 51-56.

Alagöz, S., Türkyılmaz, M., Tağı, Ş. and Özkan, M. (2015). Effects of different sorbic acid and moisture levels on chemical and microbial qualities of sun-dried apricots during storage. *Food Chem., 174*, 356-364.

Alasalvar, C. (2010). Seafood Quality, Safety and Health Applications. John Wiley and Sons. ISBN 978-1-4051-8070-2.

Alexandrou, O., de W.B. and Adams, M. R. (1995). Capacitance measurement to assess acid-induced injury to Salmonella enteritidis PT4. *Int. J. Food Microbiol., 27,* 27–36.

Anderson, M. E., Marshall, R. T. and Dickson, J. S. (1992). Efficacies of acetic, lactic and two mixed acids in reducing numbers of bacteria on surfaces of lean meat. *J. Food Saf., 12,* 139–147.

Anon (2012). Introduction to Chemistry: General, Organic, and Biological. Chapter 15: "Organic acids and bases and some of their derivatives," pp. 789-889. Available at: http://2012books.lardbucket.org/books/introduction-to-chemistry-general-organic-and-biological/ Acceseed May 2, 2016.

Anon (2016). Hurdle Technology. Available at: https://en.wikipedia.org/wiki/Hurdle_technology. Accessed March 3, 2016.

Anyasi, T. A., Jideani, A. I. O. and Mchau, G. R. A. (2015). Effect of organic acid pretreatment on some physical, functional and antioxidant properties of flour obtained from three unripe banana cultivars. *Food Chem., 172,* 515-522.

Back, S., Jin, H. and Lee, S. (2009). Inhibitory effect of organic acids against Enterobacter sakazakii in laboratory media and liquid foods. *Food Control, 20(10),* 867-872.

Barbosa-Canovas, G. V., Fernandez-Molina, J. J., Alzamora, S. M., Tapia, M. S., Lopez-Malo, A. and Chanes, J. W. (2003). General considerations for preservation of fruits and vegetables. In: Handling and Preservation of Fruits and Vegetables by Combined Methods for Rural Areas. Rome: Food and Agriculture Organization of the United Nations.

Bilau, M., Matthys, C., Vinkx, C. and de Henauw, S. (2008). Intake assessment for benzoates in different subgroups of the Flemish population. *Food Chem. Toxicol., 46(2)*, 717–723.

Birk, T., Grønlund, A.C., Christensen, B. B., Knøchel, S., Lohse, K. and Rosenquist, H. (2010). Effect of organic acids and marination ingredients on the survival of *Campylobacter jejuni* on meat. *J. Food Prot., 73,* 258–265.

Bracey, D., Holyoak, C. D. and Coote, P. J. (1998). Comparison of the inhibitory effect of sorbic acid and amphotericin B on *Saccharomyces cerevisiae*: is growth inhibition dependent on reduced intracellular pH? *J. Appl. Microbiol., 85,* 1056-1066.

Brul, S. and Coote, P. (1999). Preservative agents in foods. Mode of action and microbial resistance mechanisms. *Int. J. Food Microbiol., 50,* 1–17.

Burdock, G. A. (2004). *Fenaroli's handbook of flavor ingredients*. Boca Raton: CRC press.

C&EN (2005). Production report – Production growth in most regions. *Chem. Eng. News, 83(28),* 67–76.

CEH, (2013). Acetic acid. Chemical Economics Handbook. Available at: https://www.ihs.com/products/acetic-acid-chemical-economics-handbook.html. Accessed May 5, 2016.

Chang, J-M. and Fang, T. J. (2007). Survival of *Escherichia coli* O157:H7 and *Salmonella enterica* serovars Typhimurium in iceberg lettuce and the antimicrobial effect of rice vinegar against *E. coli* O157:H7. *Food Microbiol., 24,* 745–751.

Cherrington, C. A., Hinton, M., Mead, G. C. and Chopra, I. (1991). Organic Acids: Chemistry, antibacterial activity and practical applications. *Adv. Microb. Physiol., 32,* 87-108. DOI: 10.1016/S0065-2911(08)60006-5.

Clauden, J., Greeves, N. and Warran, S. (2012). Organic chemistry (2nd edition) New York. USA: Oxford University Press.

Crampton, L. (2016). Health Effects of Benzoic Acid, Sodium Benzoate and Benzene. Available at: https://caloriebee.com/nutrition/Effects-of-Benzoic-Acid-and-Benzoates-in-Food-and-Medicines. Accessed May 29, 2016.

Davidson, P. M. (1997). Chemical preservatives and natural antimicrobial compounds. In: M. P. Doyle, L. R. Beuchat, and T. J. Montville (Eds), *Food microbiology: fundamentals and frontiers* (pp 520-556). Washington: American Society of Microbiology.

Davidson, P. M., Sofos, J. N. and Branen, A. L. (2005). Organic acids. In: P. M. Davidson, J. N. Sofos, and A. L. Branen (Eds), *Antimicrobials in food* (3rd Edition, pp 720). CRC Press. DOI: 10.1201/9781420028737.

deMan, J. M. (1999). Principles of Food Chemistry, Third Edition. Gaithersburg, Maryland: Chapman and Hall Food Science, Aspen Publishers.

Dibner, J. J. and Buttin, P. (2002). Use of organic acids as a model to study the impact of gut microflora on nutrition and metabolism. *J. Appl. Poult. Res. 11,* 453–463.

Dishisha, T., Ibrahim, M. H., Cavero, V. H., Alvarez, M. T. and Hatti-Kaul, R. (2015). Improved propionic acid production from glycerol: Combining cyclic batch-and sequential batch fermentations with optimal nutrient composition. *Bioresource Technol., 176*, 80-87.

Doores, S. (1993). Organic acids. In: P. M. Davidson, and A. L. Branen (Eds), *Antimicrobials in foods* (2nd edn, pp 95 - 136). New York: Marcel Dekker Inc.

Doores, S. (2005). Organic acids. In: P. M. Davidson, J. N. Sofos, and A. L. Branen (Eds), *Antimicrobials in food* (3rd Edition, pp 91-142). New York: Marcel Dekker Inc.

Dubal, Z. B., Paturkar, A. M., Waskar, V. S., Zende, R. J., Latha, C., Rawool, D. B. and Kadam, M. M. (2004). Effect of food grade organic acids on inoculated *S. aureus*, *L. monocytogenes*, *E. coli* and *S.* Typhimurium in sheep/goat meat stored at refrigeration temperature. *Meat Sci, 66,* 817–821.

Eklund, T. (1985). The effect of sorbic acid and esters of *para-* hydroxyl benzoic acid on the proton motive force in *Escherichia coli* membrane vesicles. *J. Gen. Microbiol. 131,* 73–76.

Ells, T. C. and Hansen, I. T. (2006). Strain and growth temperature influence *Listeria* spp. attachment to intact and cut cabbage. *Int. J. Food Microbiol., 111,* 34–42.

EU, (2004). Regulation (EC) No 853/2004 of the European Parliament and of the Council of 29 April 2004 laying down specific hygiene rules for food of animal origin. Available at: http://www.chem.purdue.edu/gchelp/molecules/ch3co2h.html. Accessed April 23, 2016.

Fang, J., Zhang, M., Zhu, P., Ouyang, J., Gong, J., Chen, W. and Xu, F. (2015). Solubility and solution thermodynamics of sorbic acid in eight pure organic solvents. *J. Chem. Thermodyn., 85*, 202-209.

Ferrand, C., Marc, F., Fritsch, P., Cassand, P. and de Saint Blanquat, G. (2000). Mutagenicity and genotoxicity of sorbic acid–amine reaction products. *Toxicol. in Vitro, 14*(5), 423-428.

Food and Agriculture Organization of the United Nations (1995). Fruit and Vegetables processing. FAO Agricultural Services Bulletin No.119. Rome.

Gil, M. I., Selma, M. V., López-Gálvez, F. and Allende, A. (2009). Fresh- cut product sanitation and wash water disinfection: problems and solutions. *Int. J. Food Microbiol., 134,* 37 – 45.

Goddard, B. L., Mikel, W. B., Conner, D. E. and Jones, W. R. (1996). Use of organic acids to improve the chemical, physical, and microbial attributes of beef strip loins stored at –1°C for 112 days. *J. Food Prot., 59,* 849–853.

Gonzalez, M. and Gonzalez V. (2012). Organic acids. In: L. M. L. Nollet, and F. Toldra (Eds, pp. 443-466), *Food Analysis by HPLC (3rd ed.).* London: CRC Press Taylor and Francis Group.

Gonzalez, R. J., Luo, Y., Ruiz-Cruz, S. and Mcevoy, J. L. (2004). Efficacy of sanitizers to inactivate *Escherichia coli* O157:H7 on fresh-cut carrot shreds under simulated process water conditions. *J. Food Prot., 67,* 2375–2380.

Hannuksela, M. and Haahtela, T. (1987). Hypersensitivity reactions to food additives. *Allergy, 42(8),* 561–575.

Helander, I. M., von Wright, A. and Mattila-sandholm, T-M. (1997). Potential of lactic acid bacteria and novel antimicrobials against Gram-negative bacteria. *Trends Food Sci. Technol., 8,* 146–150.

Helander, I. M., von Wright, A. and Mattila-Sandholm, T-M. (1997). Potential of lactic acid bacteria and novel antimicrobials against Gram-negative bacteria. *Trends Food Sci. Technol., 8,* 146–150.

Hirshfield, I.N., Terzulli S. and Byrne, C. (2003). Weak organic acids: a panoply of effects on bacteria. *Sci. Progress, 86,* 245–269.

Holyoak, C. D., Stratford, M., McMullin, Z., Cole, M. B., Crim- mins, K., Brown, A. J. P. and Coote, P. (1996). Activity of the plasma- membrane H1-ATPase and optimal glycolytic flux are required for rapid adaptation and growth in the presence of the weak acid preservative sorbic acid. *Appl. Environ. Microbiol., 62,* 3158–3164.

http://www.newworldencyclopedia.org/entry/Acetic_acid. Accessed May 2, 2016.

https://en.wikipedia.org/wiki/Carboxylic_acid. Accessed April 22, 2016.

Huang, Y., Wilson, M., Chapman, B. and Hocking, A. D. (2010). Evaluation of the efficacy of four weak acids as antifungal preservatives in low-acid intermediate moisture model food systems. *Food Microbiol., 27,* 33–36.

Hugas, M. and Tsigarida, E. (2008). Pros and cons of carcass decontamination: the role of the European Food Safety Authority. *Meat Sci., 78,* 43–52.

Juhlin, L. (1981). Recurrent urticaria: clinical investigation of 330 patients. *Brit. J. Dermatol., 104(3),* 369–381.

Karapinar, M. and Gönül, S. A. (1992). Removal of *Yersinia enterocolitica* from fresh parsley by washing with acetic acid or vinegar. *Int. J. Food Microbiol., 16,* 261–264.

Kim, J. and Marshall, D. L. (2000). Lactic acid inactivation of *Salmonella typhimurium* attached to catfish skin. *J. Food Saf., 20,* 53–64.

Kim, J. and Marshall, D. L. (2001). Effect of lactic acid on *Listeria monocytogenes* and *Edwardsiella tarda* attached to catfish skin. *Food Microbiol., 18,* 589–596.

Kim, J. and Marshall, D. L. (2002). Influence of catfish skin mucus on trisodium phosphate inactivation of attached *Salmonella* Typhimurium, *Edwardsiella tarda*, and *Listeria monocytogenes. J. Food Prot., 65,* 1146–1151.

Koczon, P. (2009). Growth inhibition mode of action of selected benzoic acid derivatives against the yeast Pichia anomala. *J. Food Prot., 72,* 791–800.

Kondo, N., Murata, M. and Isshiki, K. (2006). Efficiency of sodium hypochlorite, fumaric acid, and mild heat in killing native microflora and *Escherichia coli* O157:H7, *Salmonella* Typhimurium DT 104, and *Staphylococcus aureus* attached to fresh-cut lettuce. *J. Food Prot., 69,* 323–329.

Koutsoumanis, K. P., Ashton, L. V., Geornaras, I., Belk, K. E., Scanga, J. A., Kendall, P. A., Smith, G. C. and Sofos, J. N. (2004). Effect of single or sequential hot water and lactic acid decontamination treatments on the survival and growth of *Listeria monocytogenes* and spoilage microflora during aerobic storage of fresh beef at 4, 10, and 25°C. *J. Food Prot., 67,* 2703–2711.

Koutsoumanis, K. and Skandamis P. (2013). New research on organic acids and pathogen behaviour. In: J. Sofos (Ed), *Advances in microbial food safety* (Vol 1, pp. 355-384). Cambridge: Woodhead Publishing Limited.

Krebs, H. A., Wiggins, D., Sole, S. and Bedoya, F. (1983). Studies on the mechanism of the antifungal action of benzoate. *Biochem. J., 214,* 657–663.

Laury, A. M., Alvarado, M. V., Nace, G., Alvarado, C. Z., Brooks, J. C., Echeverry, A. and Brashears, M. M. (2009). Validation of a lactic acid- and citric acid-based antimicrobial product for the reduction of *Escherichia coli* O157:H7 and *Salmonella* on beef tips and whole chicken carcasses. *J. Food Prot., 72,* 2208–2211.

Lee, S. (2004). Microbial safety of pickled fruits and vegetables and hurdle technology. *Internetional J. Food Saf., 4,* 21-32.

Lee, S. Y., Yu, K. M., Fellman, and Kang, D. H. (2002). Inhibition of Salmonella typhimurium and Listeria monocytogenes in mung bean sprouts by chemical treatment. *J. Food Prot., 65,* 1088–1092.

Leistner, I. (2000). Basic aspects of food preservation by hurdle technology. *Int. J. Food Microbiol., 55,* 181–186.

Leite de Souza, E., Tayse da Cruz Almeida, E. and Pereira de Sousa, G. J. (2016). The Potential of the incorporation of essential oils and their individual constituents to improve microbial safety in juices: A Review. *Compr. Rev. Food Sci. Food Saf.,* doi: 10.1111/1541-4337.12208.

Lianou, A., Koutsoumanis K. P. and Sofos J. N. (2012). Organic acids and other chemical treatments for microbial decontamination of food. In: A. Demirci, and M. O. Ngadi (Eds), *Microbial Decontamination in the Food Industry: Novel Methods and Applications.* Cambridge: Woodhead Publishing.

Liao, C.H., Shollenberger L.M. and Phillip J.G. (2003). Lethal and sublethal action of acetic acid on Salmonella in vitro and on cut surfaces of apple slices. *J. Food Sci., 68,* 2793–2798.

López-Gálvez, F., Allende, A., Selma, M. V. and Gil, M. I. (2009). Prevention of *Escherichia coli* cross-contamination by different commercial sanitizers during washing of freshcut lettuce. *Int. J. Food Microbiol.,* 133, 167–171.

Loretz, M., Stephan, R. and Zweifel, C. (2010). Antimicrobial activity of decontamination treatments for poultry carcasses: a literature survey. *Food Control, 21,* 791–804.

Loretz, M., Stephan, R. and Zweifel, C. (2011a). Antibacterial activity of decontamination treatments for cattle hides and beef carcasses. *Food Control, 22,* 347–359.

Loretz, M., Stephan, R. and Zweifel, C. (2011b). Antibacterial activity of decontamination treatments for pig carcasses. *Food Control, 22,* 1121–1125.

Luck, E. (1977). Conservacion Quimica De Los Alimentos. 1st edn. Zaragoza Editorial Acribia.

Malveda, M. (2007). Acetic acid. Chemical Economics Handbook CEH Report 602.5000, SRI International.

Mani-López, E., García H. S. and López-Malo, A. (2012). Organic acids as antimicrobials to control Salmonella in meat and poultry products. *Food Res. Int., 45,* 713–721.

Marshall, D. L. and Kim, C. R. (1996). Microbiological and sensory analyses of refrigerated catfish fillets treated with acetic and lactic acids. *J. Food Qual., 19,* 317–329.

Mastromatteo, M., Conte, A. and Del Nobile M. A. (2010). Combined Use of Modified Atmosphere Packaging and Natural Compounds for Food Preservation. *Food Eng. Rev., 2,* 28–38. DOI 10.1007/s12393-010-9013-5.

Mattey, M. and Kristiansen, B. (1999). A Brief Introduction to Citric Acid Biotechnology. In B. Kristiansen, M. Mattey, and J. Linden (Eds), *Citric Acid Biotechnology,* London: Taylor and Francis Ltd.

McMurry, J. (2008). *Carboxylic acids: Organic Chemistry* (7th Edition). Belmont, CA: Thomson Brooks/Cole, Thomson learning Inc.

Munyaka, A., Oey, I., Van Loey, A. and Hendrickx, M. (2010). Application of thermal inactivation of enzymes during vitamin C analysis to study the influence of acidification, crushing and blanching on vitamin C stability in broccoli (*Brassica oleracea* L var. italica). *Food Chem. 120,* 591-598.

Nollet, L. M. L. and Toldra, F. (2012). Food Analysis by HPLC, Third Edition, CRC Press.

Ohtsuki, T., Sato, K., Sugimoto, N., Akiyama, H. and Kawamura, Y. (2012). Absolute quantitative analysis for sorbic acid in processed foods using proton nuclear magnetic resonance spectroscopy. *Anal. Chim. Acta, 734,* 54-61.

Okolocha, E. C. and Ellerbroek, L. (2005). The influence of acid and alkaline treatments on pathogens and the shelf life of poultry meat. *Food Control, 16,* 217–225.

Olmez, H. (2010). Effect of different sanitizing methods and incubation time and temperature on inactivation of *Escherichia coli* on lettuce. *J. Food Saf., 30,* 288–299.

Olmez, H. and Kretzschmar, U. (2009). Potential alternative disinfection methods for organic fresh-cut industry for minimizing water consumption and environmental impact. *LWT-Food Sci. Technol., 42,* 686–693.

Otero-Losada, M. E. (2003). Differential changes in taste perception induced by benzoic acid pickling. *Physiol. Behav., 78,* 415–425.

Pastrorova, I., de Koster, C. G. and Boom, J. J. (1997). Analytic Study of Free and Ester Bound Benzoic and Cinnamic Acids of Gum Benzoin Resins by

GC-MS HPLC-frit FAB-MS. *Phytochem. Anal., 8(2), 63–73. doi:10.1002/(SICI)1099-1565(199703)8:2<63::AID-PCA337>3.0.CO;2-Y.*

Piper, P., Mahe, Y., Thompson, S., Pandjaitan, R., Holyoak, C., Egner, R., Muhlbauer, M., Coote, P. and Kuchler, K. (1998). The Pdr12 ATP-binding cassette ABC is required for the development of weak acid resistance in *Saccharomyces cerevisiae. EMBO J., 17,* 4257-4265.

Piper, P.W. (2011). Resistance of Yeasts to Weak Organic Acid Food Preservatives. *Adv. Appl. Microbiol., 77,* 97-113.

Podolak, R. K., Zayas, J. F., Kastner, C. L. and Fung, D. Y. C. (1996). Inhibition of *Listeria monocytogenes* and *Escherichia coli* O157:H7 on beef by application of organic acids. *J. Food Prot.,* 59, 370–373.

Poulter, S. (2007). The proof food additives are as bad as we feared. *Daily Mail Online.*

Priecina, L. and Karklina, D. (2015) Composition of major organic acids in vegetables and spices. CBU International conference on Innovation, Technology, Transfer and Education. March 25-27, 2015, Prague, Czech Republic www.cbuni.cz, ojs.journals.cz. UDC classification: 542.9, 447 454. DOI: http://dx.doi.org/10.12955/cbup.v3.637.

Quitmann, H., Fan, R. and Czermak, P. (2014). Acidic Organic Compounds in Beverage, Food, and Feed Production. *Adv. Biochem. Eng. Biotechnol. 143,* 91–141. DOI: 10.1007/10_2013_262.

Raftari, M., Jalilian F.A., Abdulamir A.S., Sekawi Z. and Fatimah A.B. (2009). Effect of organic acids on Escherichia coli O157:H7 and Staphylococcus aureus contaminated meat. *Open Microbiol. J., 3,* 121–127.

Rajkovic, A., Smigic, N. and Devlieghere, F. (2010). Contemporary strategies in combating microbial contamination in food chain. *Int. J. Food Microbiol., 141,* S29–S42.

Raybaudi-Massilia R. M., Mosqueda-Melgar J., Soliva-Fortuny R. and Martín-Belloso, O. (2009). Control of pathogenic and spoilage microorganisms in fresh- cut fruits and fruit juices by traditional and alternative natural antimicrobials. *Compr. Rev. Food Sci. Food Saf., 8,* 157 – 180.

Ricke, S. C. (2003). Perspectives of the use of organic acids and short chain fatty acids as antimicrobials. *Poult. Sci., 82,* 632–639.

Riedel, C. T., Brøndsted, L., Rosenquist, H., Haxgart, S. N. and Christensen, B. B. (2009). Chemical decontamination of *Campylobacter jejuni* on chicken skin and meat. *J. Food Prot., 72,* 1173–1180.

Salmond, C. V., Kroll, R. G. and Booth, I. R. (1984). The effect of food preservatives on pH homeostasis in Escherichia coli. *J. Gen. Microbiol. 130,* 2845–2850.

Samelis, J., Sofos J. N., Kendall, P. A. and Smith, G. C. (2001b). Fate of *Escherichia coli* O157:H7, *Salmonella* Typhimurium DT 104, and *Listeria monocytogenes* in fresh meat decontamination fluids at 4 and 10°C. *J. Food Prot., 64,* 950–957.

Santini, A. O., Pezza, H. R., Carloni-Filho, J., Sequinel, R. and Pezza, L. (2009). Potentiometric sensor for sorbic acid determination in food products. *Food Chem. 115(4),* 1563-1567.

Sapers, G. M. (2001). Efficacy of washing and sanitizing methods for disinfection of fresh fruit and vegetable products. *Food Technol. Biotechnol, 39,* 305 – 311.

Saraiva, G., Nogueira, C., Freire, P., de Sousa, F., da Silva, J., Teixeira, A. and Mendes F. J. (2015). Temperature-dependent vibrational spectroscopic study and DFT calculations of the sorbic acid. *Spectrochim. Acta A, 137*, 1409-1416.

Sengum, I. Y. and Karapinar, M. (2004). Effectiveness of lemon juice, vinegar and their mixture in the elimination of *Salmonella typhimurium* on carrots (*Daucus carota* L.). *Int. J. Food Microbiol., 96,* 301–305.

Sieber, R., Butikofer, U. and Bosset, J.O. (1995). Benzoic acid as a natural compound in cultured dairy products and cheese. *Int. Dairy J., 5,* 227–246.

Sieber, R., Bütikofer, U., Bosset, J. O. and Rüegg, M. (1989). Benzoesäure als natürlicher Bestandteil von Lebensmitteln- eine Übersicht. *Mitteilungen aus dem Gebiete der Lebensmitteluntersuchung und Hygiene*, *80,* 345–362.

Singh, N., Singh, R. K., Bhunia, A. K. and Stroshine, R. L. (2002). Effect of inoculation and washing methods on the efficacy of different sanitizers against *Escherichia coli* O157:H7 on lettuce. *Food Microbiol., 19,* 183–193.

Siragusa, G. R. (1995). The effectiveness of carcass decontamination systems for controlling the presence of pathogens on the surfaces of meat animal carcasses. *J. Food Saf., 15,* 229–238.

Skandamis, P.N., Nyogas G-J.E. and Sofos, J.N. (2010). Meat decontamination. In: F. Toldra (Ed), *Handbook of Meat Processing* (pp. 43-85) Ames, IA: Blackwell Publishing.

Smelt, J. P. P. M, (1998). Recent advances in the microbiology of high pressure processing. *Trends Food Sci. Technol., 9,* 152-158.

Smith, J. and Hong-Shum, L. (2011). Food additives data book. Oxford: Wiley.

Soomro, A. H., Masud T. and Anwaar, K. (2002). Role of Lactic Acid Bacteria (LAB) in Food Preservation and Human Health – A Review. *Pak. J. Nutr. 1(1),* 20-24.

Steeg, E. and Montag, A. (1987). Nachweis aromatischer Carbonsäuren in Honig. *Zeitschrift für Lebensmittel-Untersuchung und - Forschung, 184,* 17–19.

Stratford, M. and Anslow, P. A. (1998). Evidence that sorbic acid does not inhibit yeast as a classic ''weak acid'' preservative. *Lett. Appl. Microbiol., 27,* 203–206.

Stratford, M. and Eklund, T. (2003). Organic acids and esters. In: N. J. Russell, and G. W. Gould (Eds), *Food preservatives* (2nd edn, pp. 48–84) New York: Kluwer Academic/Plenum Publishers.

Stratford, M., Plumridge, A., Pleasants, M. W., Novodvorska, M., Baker-Glenn, C. A., Pattenden, G. and Archer, D. B. (2012). Mapping the structural requirements of inducers and substrates for decarboxylation of weak acid preservatives by the food spoilage mould aspergillus niger. *Int. J. Food Microbiol., 157(3),* 375-383.

Suhr, K. I. and Nielsen, P. V. (2004). Effect of weak acid preservatives on growth of bakery product spoilage fungi at different water activities and pH values. *Int. J. Food Microbiol., 95,* 67-78.

Ter Beek, A., Wijman, J. G., Zakrzewska, A., Orij, R., Smits, G. J. and Brul, S. (2015). Comparative physiological and transcriptional analysis of weak organic acid stress in bacillus subtilis. *Food Microbiol., 45*, 71-82.

Tfouni, S. A. V. and Toledo, M. C. F. (2002) Determination of benzoic and sorbic acids in Brazilian food. *Food Control, 13,* 117–123.

Theron, M. M. and Lues, J. F. R. (2007). Organic acid and meat preservation: a review. *Food Rev. Int., 23,* 141-158.

Theron, M. M. and Lues, J. F. R. (2011). *Organic acids and food preservation.* Boca Raton, FL: CRC Press.

Theron, M. M. and Rykers, J. F. (2010). *Nature and composition of organic acids: Organic acids and food preservation.* New York: CRC Press Taylor and Francis Group.

Ukuku, D. O. and Sapers, G. M. (2001). Effect of sanitizer treatments on *Salmonella* Stanley attached to the surface of cantaloupe and cell transfer to fresh-cut tissues during cutting practices. *J. Food Prot., 64,* 1286–1291.

Ukuku, D. O., Pilizota, V. and Sapers, G. M. (2001). Influence of washing treatments on native microflora and *Escherichia coli* population of inoculated cantaloupes. *J Food Saf, 21,* 31–47.

USDA-FSIS (2010). Safe and suitable ingredients used in the production of meat and poultry products. Directive 7120.1, Rev. 2. www.fsis.usda.gov/OPPDE/rdad/FSISDirectives/7120.1Rev2.pdf Accessed January 11, 2011.

Van Netten, P., Huisin't Veld J.H. and Mossel D.A. (1994). The immediate bactericidal effect of lactic acid on meat-borne pathogens. *J. Appl. Bacteriol., 77,* 490–496.

Vijayakumar, C. and Wolf-Hall, C. E. (2002). Evaluation of household sanitizers for reducing levels of *Escherichia coli* on iceberg lettuce. *J. Food Prot., 65,* 1646–1650.

Warth, A. D. (1991). Mechanism of action of benzoic acid on Zygosaccharomyces bailii: effects on glycolytic metabolite levels, energy production, and intracellular pH. *Appl. Environ. Microbiol., 57(12),* 3410–4.

World Health Organization (WHO). (2000). Benzoic acid and sodium benzoate. *Concise International Chemical Assessment Document Number* 26. Geneva (Switzerland): WHO.

Wu, F. M., Doyle, M. P., Beuchat, L. R., Wells, J. G., Mintz, E. D. and Swaminathan, B. (2000). Fate of *Shigella sonnei* on parsley and methods of disinfection. *J. Food Prot., 63,* 568–572.

Yasothai, R. and Giriprasad, R. (2015a). Weak organic acids in Food Technology. *International J. Sci. Environ. Technol., 4(1),* 164 – 166.

Yasothai, R. and Giriprasad, R. (2015b). High pressure processing food technologies. *Int. J. Sci. Environ. Technol., 4(1),* 108 – 113.

Yoneda, N., Kusano, S., Yasui, M., Pujado, P. and Wilcher, S. (2001*). Appl. Catal. A-Gen., 221,* 253–265.

In: Organic Acids
Editor: Cesar Vargas

ISBN: 978-1-63485-931-8

Chapter 2

CHROMATOGRAPHIC ANALYSIS OF ORGANIC ACIDS IN FOOD FROM ANIMAL ORIGIN

Marion P. Costa[1,2] ***and Carlos A. Conte-Junior***[1,2,*]

[1]Universidade Federal Fluminense, Faculdade de Veterinária, Department of Food Technology, Niterói, RJ, Brazil

[2]Universidade Federal do Rio de Janeiro, Instituto de Química, Food Science Program, Rio de Janeiro, RJ, Brazil

ABSTRACT

Organic acids are compounds with acidic properties and occur naturally in a number of foods. They are mainly present in fermented products as a result of hydrolysis, biochemical metabolism, and microbial activity. Even so these are not considered as nutrients, however, they are responsible for giving a characteristic taste to food. In addition, the organic acids have been widely used as food additives and preservatives for avoiding food deterioration and extending the shelf life of different products. For these reasons, determining organic acid content in food products is important, since these compounds contribute to the flavor and aromatic properties of them. Besides, organic acids can influence the preservation of some foods. However, they are not members of a

[*] Corresponding author: Carlos A. Conte-Junior, M.Sc., Ph.D.; Rua Vital Brazil Filho, n. 64. Santa Rosa; CEP: 24.230-340; Niterói, Rio de Janeiro, Brazil; Phone: +55 21 – 2629-9545; E-mail address: carlosconte@id.uff.br (C. A. Conte-Junior).

homologous series, which differ in the number of carboxy groups, hydroxy groups, and carbon–carbon double bonds in their molecules. The lowest monocarboxylic aliphatic acids, such as formic, acetic, propionic, and butyric, are rather volatile liquids, whereas those acids containing more carbon atoms are of a relatively oily substance and slightly water soluble. Alicyclic acids are less water soluble than the previous ones. In comparison, dicarboxylic acids are colorless solids with melting points at about 100°C. All these acids form somewhat soluble metal salts and esters, of which the latter are adequately volatile for gas chromatography (GC) analysis using flame ionization or mass spectrometric (MS) detectors. Besides that, they also have spectral absorption properties that make them suitable for high-performance liquid chromatography (HPLC) analysis using ultraviolet, refractive index or MS detectors. In this context, GC has been used to determine the volatile organic acids, while HPLC has been widely used for analyzing non-volatile organic acids in different matrixes. There are some studies evaluating organic acids in honey, dairy, fish and meat products, however, the best tool to each food matrix depend on of the ingredients and the technological process applied by the food industry.

INTRODUCTION

Organic acids are organic compounds with acidic properties characterized by a carboxyl group (-COOH), which in its chemical structure are composed of carbon. These compounds are classified according to: the type of carbon chain (aliphatic, alicyclic, aromatic and heterocyclic); the extent of unsaturation (saturated and unsaturated); and the number of functional groups (monocarboxylic, dicarboxylic etc.) (Quitmann, Fan, and Czermak, 2013). These differences in chemical structure are responsible for distinct characteristics of organic acids. For example, monocarboxylic aliphatic acid with up to four carbon atoms is highly volatile liquids, whereas those with five or more carbon atoms are slightly water-soluble liquids (Käkölä, Alén, Pakkanen, Matilainen, and Lahti, 2007).

The organic acids are responsible for giving a characteristic taste of different types of foods, such as yogurts and fermented meat (Costa and Conte-Junior, 2015). However, these are not considered as nutrients. These compounds occur naturally in a number of foods, mainly in fermented products as a result of hydrolysis, biochemical metabolism and microbial activity (Swetwiwathana and Visessanguan, 2015). Moreover, the organic acids have been widely used for the food industry as food additives and preservatives for avoiding food deterioration and extending the shelf life of food ingredients (Cheng, 2010; Jurado-Sánchez, Ballesteros, and Gallego, 2011) due antimicrobial activity (Mani-López, García, and López-Malo, 2012; Mohan and Pohlman, 2016; Zaki, Mohamed, and El-Sherif, 2015).

For these reasons, determining organic acid content in food products is important. Since they contribute to the flavor and aromatic properties of them (Farajzadeh and Assadi, 2009; Kritsunankul, Pramote, and Jakmunee, 2009; Tormo and Izco, 2004) and to food preservation (Cruz-Romero, Murphy, Morris, Cummins, and Kerry, 2013; Mani-López et al., 2012). Their presence and the relative ratio of organic acids can affect the chemical and sensorial characteristics of the food matrix (e.g., pH, total acidity and microbial stability) and can provide information on nutritional properties of food and means to optimize selected technological processes (Chinnici, Spinabelli, Riponi, and Amati, 2005). The quantitative determination of organic acids is also important to monitor bacterial growth and metabolic activity (Costa, Silva Frasao, Costa Lima, Rodrigues, and Conte-Junior, 2016). In this context, the high-performance liquid chromatography has been widely used for analyzing non-volatile organic acids in complex matrixes such as yogurt, cheese, and meat products. Besides, the gas chromatography can be used to determine the volatile organic acids in some matrix.

Organic Acids in Food from Animal Origin

Organic acids in foods of animal origin mainly result from the metabolism of large-molecular-mass compounds, such as carbohydrates, lipids, and proteins. These acids are also found in many products as compounds added to food to carry out some hygienic or technologic function (Brul and Coote, 1999). Organic acids such as lactic and acetic acids are used as direct antimicrobial activity products and are incorporated into human foods (Cruz-Romero et al., 2013), because of their ability to lower the pH, resulting in instability of bacterial cell membranes (Mani-López et al., 2012). These acids can accumulate over time as they are produced by fermentation activity of indigenous microorganisms starter cultures or added (Costa and Conte-Junior, 2013; Ricke, 2003).

In milk, the organic acid content varies from 0.12% to 0.21%, or around 1.2% dry matter. The citric acid is the predominant organic acid in milk normally present in the form of citrate (Walstra, 2013). During storage, citric acid disappears rapidly as a result of bacterial growth. Lactic and acetic acids are degradation products of lactose. Other acids are produced from the hydrolysis of lactose, citric acid, and lipid (Leite et al., 2013). Milk also contains nitrogenous acidic compounds such as orotic acid and hippuric acid. The orotic acid concentration is mainly influenced by diet and stage of lactation (Tormo and Izco, 2004). In fermented milk, generally, the production of some organic acids, such as lactic, formic, acetic, and succinic, is the result of the metabolic activity of the starter cultures (Ammor, Tauveron, Dufour, and Chevallier, 2006). In natural yogurt, for example, the lactic acid is the predominant organic acid (Cutrim et al., 2016; Costa et al., 2016). While in cheese ripening, the products of primary events such as free fatty acids, organic acids, and free amino acids are further catabolized to smaller volatile and nonvolatile flavor compounds (Subramanian, Alvarez, Harper, and Rodriguez-Saona, 2011; Suomalainen and Mäyrä-Makinen, 1999). Thus, the organic acids present in the various types of cheese can vary according to the manufacturing process and cheese starter culture (Dimitrellou, Kandylis, Kourkoutas, Koutinas, and Kanellaki, 2015).

The predominant acid in muscle tissue is the lactic acid formed by glycolysis during *post-mortem*, followed by glycolic and succinic acids. Pyruvate, generated as the end product of glycolysis, is converted to lactic acid by a lactic dehydrogenase. Since the metabolic waste products cannot be removed without blood flow, the lactic acid accumulates in the muscle. Other acids of the Krebs cycle are present in negligible amounts (Greaser, 2001; Koohmaraie and Geesink, 2006; Kristoffersen, Tobiassen, Steinsund, and Olsen, 2006). Furthermore, lactic and acetic acids may be present in meat because they are used in the beef industry to decontaminate carcasses or meat products. The effectiveness of these acids depends on the concentration and temperature of the acid solution, exposure time, application pressure, stage in the slaughtering process, tissue type, group of microorganisms, and initial concentration (Li, Kundu, and Holley, 2015). Therefore, a higher concentration of lactic and/or acetic acid might be expected in meats treated with these acids (Carpenter, Smith, and Broadbent, 2011). In fermented meat products, the production of organic acids by bacteria is undoubtedly the determining factor for the shelf life and safety of the final product (Maijala, Eerola, Aho, and Hirn, 1993). Several factors can affect the type of organic acid present, including the microorganism involved in the fermentation process. However, few studies have assessed the production of organic acids in meat products.

Lactic acid is also the main organic acid in fish meat. During the storage of fish, some organic acids are formed include formic, acetic, propionic, n-butyric, isobutyric, n-valeric, and isovaleric acids (Osako et al., 2005). As they are for animal meats, organic acids are also used as additives for the conservation of fish and derivatives (Calo-Mata et al., 2008; García-Soto, Fernández-No, Barros-Velázquez, and Aubourg, 2014; Mejlholm et al., 2010; Mejlholm and Dalgaard, 2007). The fermentation process of fish products is similar to that at fermented meat, with lactic acid as the major product. In their study of Thai fermented fish under 4 different treatments (natural fermentation; inoculated with *Lactobacillus plantarum* IFRPD P15; inoculated with *L. reuteri* IFRPD P17; and inoculated with mixed starter culture (*L. plantarum* IFRPD P15 × *L. reuteri* IFRPD P17), (Saithong, Panthavee, Boonyaratanakornkit, and Sikkhamondhol, 2010) evaluated the production of 5 organic acids (lactic, acetic, butyric, propionic, and gluconic). They observed that lactic and gluconic acids were present in all treatments, but their behavior differed depending on the treatment. Butyric, succinic, acetic, and propionic acids were not detected in any treatment during fermentation. There is a lack of information about organic acids in the meat of different fish species and their derived products.

Honey acidity is mainly due to the presence of organic acids. The acidity contributes to the flavor, stability in the presence of microorganisms, enhancement of chemical reactions, and antibacterial and antioxidant activities. Gluconic acid, resulting from the action of honey's glucose oxidase on glucose, contributes most to the acidity and is in equilibrium with gluconolactone. Other acids, such as acetic, butyric, lactic, citric, succinic, formic, malic, maleic, and oxalic acids, are also present in small amounts. The organic acids together with inorganic anions, also contribute to the acidity of honey. The organic acids comprise a small proportion of honey (0.5%) and together with the total acidity can be used as an indicator of deterioration due to storage or aging, or to measure the purity and authenticity (Cavia, Fernández-Muiño, Alonso-Torre, Huidobro, and Sancho, 2007). They are also components of the honey flavor. Some organic acids identified in honey may be useful for characterizing different honey types. For example, the citric acid concentration is used as a reliable parameter for the differentiation of 2 main types of honey, floral and honeydew (Daniele, Maitre, and Casabianca, 2012).

HIGH-PERFORMANCE LIQUID CHROMATOGRAPHY

The analysis of organic acids in different food, such as dairy products, meat products, and honey is of great interest for the food industry. These compounds are responsible for sensory properties and food preservation (Theron and Lues, 2010). For this reason, different HPLC techniques have been used for the separation and identification of organic acids in different foods (van Hees, Dahlén, Lundström, Borén, and Allard, 1999), such as those of animal origin (Costa and Conte-Junior, 2015; Costa et al., 2016). HPLC methods have gained importance in these analyses because of the speed, selectivity, sensitivity, and reliability of this technology (Chen, Mowery, Castleberry, van Walsum, and Chambliss, 2006).

Sample Preparation

For organic acids analysis, the sample preparation step is usually simple. The extraction is preferably performed using an acid, which may be the same of the mobile phase, but with a higher concentration, such as sulfuric and phosphoric acids, or with distilled water. However, for meat samples, perchloric acid (PCA) is the most often used and the most efficient (Costa and Conte-Junior, 2015). After extraction, a centrifugation step may be used, depending mainly on the type of food to be analyzed. Most investigators who apply centrifugation use a force range from 6000 to 17000 × *g*; however, in dairy products, the use of 5500 × *g* of rotation is sufficient (Costa et al., 2016). The use of centrifugation in the analysis of organic acids in complex matrices facilitates the extraction, yielding a purer final extract. The supernatant generally is filtered through a 0.22- or 0.45-µm cellulose acetate filter, and the preparation obtained is then ready to inject into the HPLC system (González de Llano, Rodriguez, and Cuesta, 1996; Kaminarides, Stamou, and Massouras, 2007; Leite et al., 2013; Suárez-Luque, Mato, Huidobro, and Simal-Lozano, 2002).

Derivatization Techniques

As a rule, pre- and post-column derivatization processes used in liquid chromatography are intended to improve the identification and detection of the different solutes to be determined. For organic acids analysis, these processes can be mandatory in the determination of carboxylic acids on account of their structural similarities and the typically low molar extinction coefficients of their chronophers, on which their sensing usually relies, owing to the lack of more sensitive properties, as fluorescence and electrochemical activity.

However, for the major organic acids, the derivatization step is not be used, when these organic acids are analyzed by ultraviolet. Since these compounds absorb energy at a wavelength of 206-220 nm (Ahmed et al., 2015; Costa and Conte-Junior, 2015; Cutrim et al., 2016; Leite et al., 2013; Murtaza et al., 2012).

Sample Separation

Liquid chromatography has simplified the analysis for various food constituents, including organic acids. In chromatography, the selection of the stationary phase is essential in order to achieve a suitable separation. A number of different separation mechanisms have been widely employed in a different matrix, which including ion-exchange, ion-exclusion, ion-par, and reverse-phase. Consequently, the choice of method in each case is dictated essentially by the types of acid to be determined and their proportions as well as by the nature of food matrix (Quirós, Lage-Yusty, and López-Hernández, 2009). For determination of organic acids in foods from animal origin, the most usual method is ion exchange chromatography followed by reverse-phase chromatography (Costa and Conte-Junior, 2015).

The ready ionization of organic acids has long been exploited for their isolation by ion-exchange chromatography, which involves the use of an ion-exchange resin as the stationary phase. This separation technique is extremely used nowadays, and the column most frequently used for this purpose is the Aminex HPX-87H 300 x 7.8 mm model from Biorad Laboratories (Adhikari, Grün, Mustapha, and Fernando, 2002; Cutrim et al., 2016; Costa et al., 2016; Donkor, Nilmini, Stolic, Vasiljevic, and Shah, 2007; Fernandez-Garcia and McGregor, 1994; González de Llano et al., 1996; Kaminarides et al., 2007; Leite et al., 2013; Madureira et al., 2012; Ong, Henriksson, and Shah, 2006; Sriphochanart and Skolpap, 2011; Zeppa, Conterno, and Gerbi, 2001). The main advantage of using this type of column is that it enables the simultaneous analysis of carbohydrates and organic acids (Figure 1) (Costa et al., 2016).

The most stationary phases used in bonded-phase chromatography in its reversed-phase mode are based on octyl (C8 columns) and octadecyl (C18 columns) functionality. The difference between the two columns will be in the length of the carbon-chain attached to the silica surface, as for organic acid analysis to C18 column is the most used (Bensmira and Jiang, 2011; Costa and Conte-Junior, 2015; Murtaza et al., 2012; Saithong et al., 2010).

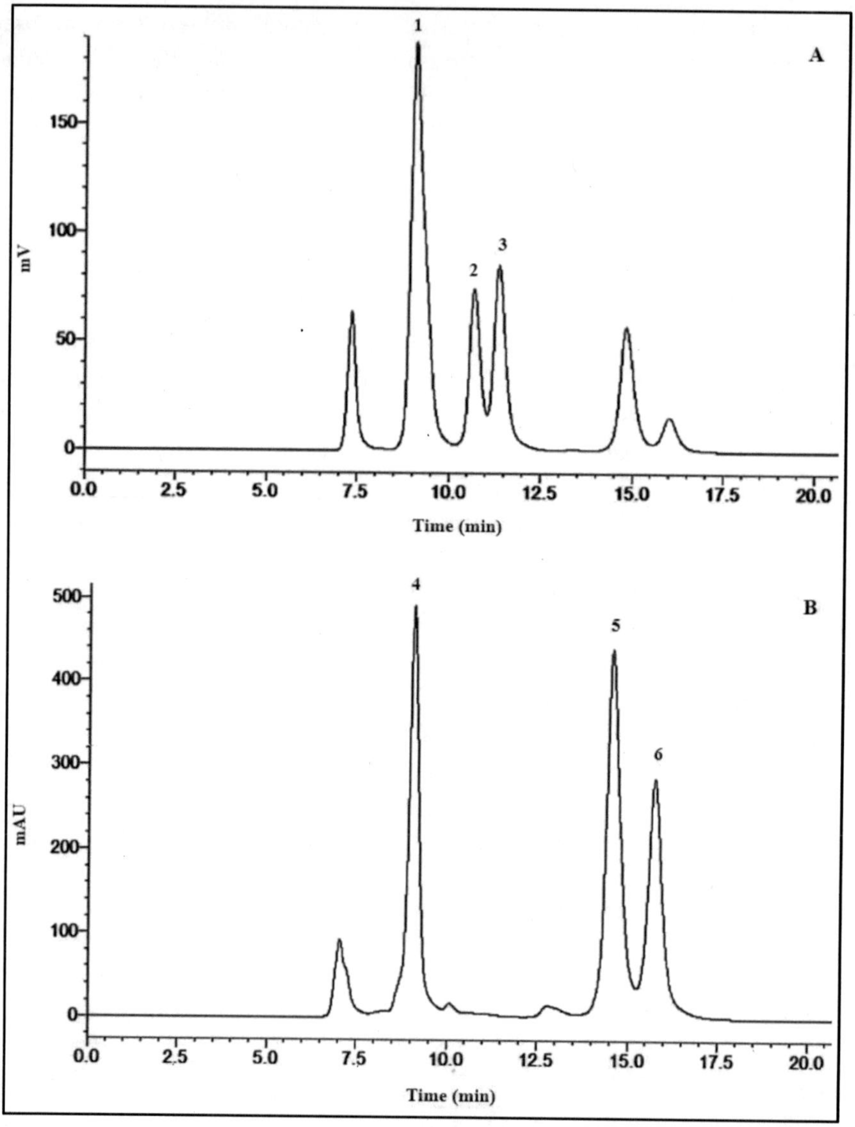

Figure 1. Chromatogram of simultaneous analysis of carbohydrates (A1) and organic acids (B1) by HPLC-DAD-RI. Lactose (1), glucose (2), galactose (3), citric acid (4), lactic acid (5), and formic acid (6).

Detection

The detectors most frequently used in HPLC for analysis of organic acids are the conductivity, the refractive index (RI) and the ultraviolet (UV), beyond mass spectrometric (MS). Nowadays, the high-performance liquid chromatography has been widely used with detection mode dual UV-VIS detector and refractive index detector for analyzing carbohydrates and non-volatile organic acids in complex matrixes, in the same chromatographic run.

The conductivity detectors were originally employed in ion chromatography for determination of inorganic ions, later for organic acids. However, the inherent difficulties have deterred potentials user from applying them to food analyses. Because this type of detector has low selectivity, and the solute conductivity measurements require the prior elimination of the eluent background conductivity using a conventional suppressing column or a more modern alternative such as a cation-exchange membrane. Currently, due to their limitations, this type of detector is not widely used (González et al., 2014).

The refractive index (RI) detector responds to a difference in the refractive index of the column effluent as it passes through the detector flow cell. For this reason, RI detection has been used very successfully for the analysis of sugars, triglycerides, and organic acids (Swartz, 2010). The RI detector is a bulk-property detector that responds to all solutes if the refractive index of the solute is sufficiently different from that of the mobile phase. These detectors are somewhat sensitive to changes in pressure, temperature, and composition of the mobile phase, this must demand strict control of the chromatographic conditions and the use of isocratic elution. However, despite its limitations RI detector has an advantage of this detectors, they can use for determining other components interest as carbohydrates, simultaneously in a single chromatographic analysis (Costa and Conte-Junior, 2015).

The most widely used detectors in modern HPLC are photometers based on ultraviolet (UV) and visible light (VIS) absorption. They have a high sensitivity for many solutes, including organic acids, but samples must absorb in the UV region (Swartz, 2010). Theses detectors are no doubt the most frequently used at present for determining organic acids in food. And they can be used for analysis of underivatized organic acids, detection at 206-220 nm, usually poses no serious problem in the determination of major organic acids (Cutrim et al., 2016; Costa et al., 2016; Donkor, Nilmini, Stolic, Vasiljevic, and Shah, 2007; Fernandez-Garcia and McGregor, 1994; González de Llano et al., 1996; Kaminarides et al., 2007; Leite et al., 2013; Madureira et al., 2012; Sriphochanart and Skolpap, 2011; Zeppa, Conterno, and Gerbi, 2001).

The mass spectrometric detector is the most sophisticated hyphenated (refer to the coupling of an independent analytical instrument to provide detection) HPLC detector in use today. In complex samples mass spectrometry coupled to liquid chromatography constitutes a powerful technique due to its high sensitivity and selectivity (Cheng, 2010; Chen et al., 2006).

Gas Chromatography

The gas chromatography (GC) is an attractive alternative to analyzing organic acids due to its simplicity, separation efficiency and excellent sensitivity and selectivity (Horák et al., 2009; Horák, Čulík, Jurková, Čejka, and Kellner, 2008; M.-H. Yang and Choong, 2001). Many short-chain organic acids are thermostable and sufficiently volatile, thus fulfilling key requirements for GC measurement. Furthermore, the method of choice for volatile acids analysis is gas chromatography is, instead of the isolation of compounds from the food matrix can be carried out by different methods, such as high vacuum distillation, simultaneous distillation extraction, supercritical fluid extraction or headspace techniques (Fernández-García, Carbonell, and Nuñez, 2002).

Sample Preparation

In general, the great complexity of food samples demands an appropriate sample preparation technique before GC analysis. As a rule, beverages usually implicate in a simple pretreatment such as dilution and/or filtration, however, for other food the potential interference of matrix compounds (e.g., lipids, vitamins, proteins, polysaccharides) require the employment of more complex pre-treatment and clean-up procedures (Kritsunankul et al., 2009).

Traditional methods such as stream distillation and liquid–liquid extraction are time-consuming and environmentally unfriendly. The solid-phase extraction (SPE) can be implemented via flow systems, resulting in a dramatically increased the process and reduced analytical cost through decreased reagent consumption (Cherchi, Spanedda, Tuberoso, and Cabras, 1994; Horák et al., 2009), other alternatives such as single-drop microextraction (Saraji and Mirmahdieh, 2009; Saraji and Mousavinia, 2006), solid-phase microextraction (Wen, Wang, and Feng, 2007) and stir-bar sorptive extraction (Horák et al., 2008) have also been successfully applied to the analysis of short and medium-chain fatty acids and preservatives in vinegar, beverages, and dairy products.

Derivatization

Although, other acids should be derivatized to convert these compounds into less polar and stable derivates suitable for their GC determination (Horák et al., 2009) to some organic acid this process is not necessary because they have a polar characteristic. To avoid the derivatization process of organic acids, there also are successfully employed capillary GC columns coated with polar stationary phases such polyethylene glycol or nitroterephthalic acid modified polyethylene glycol. When using these columns, it is possible to obtain a good chromatographic resolution, avoiding peak tailing (Horák et al., 2008; M.-H. Yang and Choong, 2001).

Detection

The flame ionization detector (FID) is the most widely applied gas chromatographic detector for hydrocarbons such as volatile organic acids, butane or hexane. With a linear range for 6 or 7 orders of magnitude (106 to 107) and limits of detection in the low picogram or femtogram range. However, the presence of oxygen molecules decreases the detector's response. Therefore, highly oxygenated molecules or sulfides might best be detected using another detector instead of the FID. Sulfides determination by the flame photometric detector (FPD) and aldehydes and ketones analyzed with the photoionization detector (PID) are alternatives to the use of the FID for those molecules (Grob and Barry, 2004).

In order to measure the characteristics of individual organic acids, mainly minority molecules in food matrix, a mass spectrometry (MS) converts them to ions so that they can be moved about and manipulated by external electric and magnetic fields. MS has been applied in food chemistry fields for the analysis of toxic compounds and contaminants, for nutraceuticals and for the characterization of foodstuff to be applied to production areas and traceability (Yang and Caprioli, 2011; Yang and Choong, 2001). Hidalgo, Navarro, Delgado, and Zamora (2013) successfully use this methodology (GC-MS) to determinate and identify eight a-keto acids (a-ketoglutaric acid, pyruvic acid, 4-hydroxyphenylpyruvic acid, 3-methyl-2-oxobutyric acid, a-keto-cmethylthiobutyric acid, 4-methyl-2-oxovaleric acid, 3-methyl-2-oxovaleric acid, and phenylpyruvic acid) in pork meat and Iberian ham samples. Therefore, MS is, today, usually coupled to HPLC or GC (Brent, Reiner, Dickerson, and Sander, 2014).

Conclusion

In this chapter, it could be evidenced that the organic acids are straight related to the intrinsic characteristic of each food from animal origin, the processing steps that these foods are submitted and the biochemical changes that occur during storage of those products. Furthermore, there are various chromatographic techniques that can be applied for the analysis of organic acids in the food matrix, appears to be the HPLC method of choice due to the chemical structure of these compounds, whereas the GC can mainly be used for identification and quantification of volatile organic acids.

Acknowledgments

The authors wish to thank the Fundação de Amparo à Pesquisa do Estado do Rio de Janeiro (process no. E-26/201.185/2014 and E-26/010.001.911/2015, FAPERJ, Brazil) and the Conselho Nacional de Desenvolvimento Científico e Tecnológico (process no. 311361/2013-7, CNPq, Brazil), and Coordenação de Aperfeiçoamento de Pessoal de Nível Superior (process no. 125, CAPES/Embrapa 2014, CAPES, Brazil) for their financial support.

References

Adhikari, K., Grün, I. U., Mustapha, A. and Fernando, L. N. (2002). Changes in the Profile of Organic Acids in Plain Set and Stirred Yogurts during Manufacture And Refrigerated Storage1. *Journal of Food Quality*, *25*(5), 435–451.

Ahmed, O. M., Pangloli, P., Hwang, C.-A., Zivanovic, S., Wu, T., D'Souza, D. and Draughon, F. A. (2015). The occurrence of Listeria monocytogenes in retail ready-to-eat meat and poultry products related to the levels of acetate and lactate in the products. *Food Control*, *52*, 43–48. http://doi.org/10.1016/j.foodcont.2014.12.015.

Ammor, S., Tauveron, G., Dufour, E. and Chevallier, I. (2006). Antibacterial activity of lactic acid bacteria against spoilage and pathogenic bacteria isolated from the same meat small-scale facility: 1—Screening and characterization of the antibacterial compounds. *Food Control*, *17*(6), 454–461. http://doi.org/10.1016/j.foodcont.2005.02.006.

Bensmira, M. and Jiang, B. (2011). Organic acids formation during the production of a novel peanutmilk kefir beverage. *British Journal of Dairy Sciences*, *2*(1), 18–22.

Brent, L. C., Reiner, J. L., Dickerson, R. R. and Sander, L. C. (2014). Method for characterization of low molecular weight organic acids in atmospheric aerosols using ion chromatography mass spectrometry. *Analytical Chemistry*, *86*(15), 7328–7336.

Brul, S. and Coote, P. (1999). Preservative agents in foods: mode of action and microbial resistance mechanisms. *International Journal of Food Microbiology*, *50*(1), 1–17.

Calo-Mata, P., Arlindo, S., Boehme, K., Miguel, T. de, Pascoal, A. and Barros-Velazquez, J. (2008). Current Applications and Future Trends of Lactic Acid Bacteria and their Bacteriocins for the Biopreservation of Aquatic Food Products. *Food and Bioprocess Technology*, *1*(1), 43–63. http://doi.org/10.1007/s11947-007-0021-2.

Carpenter, C. E., Smith, J. V. and Broadbent, J. R. (2011). Efficacy of washing meat surfaces with 2% levulinic, acetic, or lactic acid for pathogen decontamination and residual growth inhibition. *Meat Science*, *88*(2), 256–260.

Cavia, M. M., Fernández-Muiño, M. A., Alonso-Torre, S. R., Huidobro, J. F. and Sancho, M. T. (2007). Evolution of acidity of honeys from continental climates: Influence of induced granulation. *Food Chemistry*, *100*(4), 1728–1733.

Cheng, H. (2010). Volatile flavor compounds in yogurt: a review. *Critical Reviews in Food Science and Nutrition*, *50*(10), 938–950.

Chen, S.-F., Mowery, R. A., Castleberry, V. A., van Walsum, G. P. and Chambliss, C. K. (2006). High-performance liquid chromatography method for simultaneous determination of aliphatic acid, aromatic acid and neutral degradation products in biomass pretreatment hydrolysates. *Journal of Chromatography A*, *1104*(1), 54–61.

Cherchi, A., Spanedda, L., Tuberoso, C. and Cabras, P. (1994). Solid-phase extraction and high-performance liquid chromatographic determination of organic acids in honey. *Journal of Chromatography A*, *669*(1), 59–64.

Chinnici, F., Spinabelli, U., Riponi, C. and Amati, A. (2005). Optimization of the determination of organic acids and sugars in fruit juices by ion-exclusion liquid chromatography. *Journal of Food Composition and Analysis*, *18*(2-3), 121–130. http://doi.org/10.1016/j.jfca.2004.01.005.

Costa, M. and Conte-Junior, C. (2013). Leites Fermentados como alimentos funcionais. *Animal Business*, *9*, 61–65.

Costa, M. and Conte-Junior, C. (2015). Chromatographic Methods for the Determination of Carbohydrates and Organic Acids in Foods of Animal Origin. *Comprehensive Reviews in Food Science and Food Safety*, *14*(5), 586–600. http://doi.org/10.1111/1541-4337.312148.

Costa, M. P., da Silva Frasao, B., da Costa Lima, B. R. C., Rodrigues, B. L. and Junior, C. A. C. (2016). Simultaneous analysis of carbohydrates and organic acids by HPLC-DAD-RI for monitoring goat's milk yogurts fermentation. *Talanta*, *152*, 162–170.

Cruz-Romero, M. C., Murphy, T., Morris, M., Cummins, E. and Kerry, J. P. (2013). Antimicrobial activity of chitosan, organic acids and nano-sized solubilisates for potential use in smart antimicrobially-active packaging for potential food applications. *Food Control*, *34*(2), 393–397.

Cutrim, C. S., de Barros, R. F., da Costa, M. P., Franco, R. M., Conte-Junior, C. A. and Cortez, M. A. S. (2016). Survival of Escherichia coli O157: H7 during manufacture and storage of traditional and low lactose yogurt. *LWT-Food Science and Technology*, *70*, 178–184.

Daniele, G., Maitre, D. and Casabianca, H. (2012). Identification, quantification and carbon stable isotopes determinations of organic acids in monofloral honeys. A powerful tool for botanical and authenticity control. *Rapid Communications in Mass Spectrometry*, *26*(17), 1993–1998. http://doi.org/10.1002/rcm.6310.

Dimitrellou, D., Kandylis, P., Kourkoutas, Y., Koutinas, A. A. and Kanellaki, M. (2015). Cheese Production Using Kefir Culture Entrapped in Milk Proteins. *Applied Biochemistry and Biotechnology*, *176*(1), 213–230.

Donkor, O. N., Nilmini, S. L. I., Stolic, P., Vasiljevic, T. and Shah, N. P. (2007). Survival and activity of selected probiotic organisms in set-type yoghurt during cold storage. *International Dairy Journal*, *17*(6), 657–665. http://doi.org/10.1016/j.idairyj.2006.08.006.

Farajzadeh, M. A. and Assadi, A. (2009). Liquid–gas–liquid technique for microextraction and preconcentration of short chain fatty acids from aqueous samples. *Journal of Separation Science*, *32*(7), 1027–1035.

Fernández-García, E., Carbonell, M. and Nuñez, M. (2002). Volatile fraction and sensory characteristics of Manchego cheese. 1. Comparison of raw and pasteurized milk cheese. *Journal of Dairy Research*, *69*(04), 579–593.

Fernandez-Garcia, E. and McGregor, J. U. (1994). Determination of organic acids during the fermentation and cold storage of yogurt. *Journal of Dairy Science*, *77*(10), 2934–2939.

García-Soto, B., Fernández-No, I. C., Barros-Velázquez, J. and Aubourg, S. P. (2014). Use of citric and lactic acids in ice to enhance quality of two fish species during on-board chilled storage. *International Journal of Refrigeration*, *40*, 390–397. http://doi.org/10.1016/j.ijrefrig.2013.12.010.

González de Llano, D., Rodriguez, A. and Cuesta, P. (1996). Effect of lactic starter cultures on the organic acid composition of milk and cheese during ripening—analysis by HPLC. *Journal of Applied Microbiology*, *80*(5), 570–276.

González, O., Blanco, M. E., Iriarte, G., Bartolomé, L., Maguregui, M. I. and Alonso, R. M. (2014). Bioanalytical chromatographic method validation according to current regulations, with a special focus on the non-well defined parameters limit of quantification, robustness and matrix effect. *Journal of Chromatography A*, *1353*, 10–27.

Greaser, M. L. (2001). Postmortem muscle chemistry. *Meat Science and Applications*, 21–37.

Grob, R. L. and Barry, E. F. (2004). *Modern practice of gas chromatography*. John Wiley and Sons. Retrieved from https://books.google.com.br/books?hl=pt-BR&lr=&id=2L_H2l5STa8C&oi=fnd&pg=PR7&dq=Grob+and+Barry+GC&ots=uS_JhbY8DT&sig=qbM-OxFJcgK4f-6RTjCHk-3xRcg.

Hidalgo, F. J., Navarro, J. L., Delgado, R. M. and Zamora, R. (2013). Determination of α-keto acids in pork meat and Iberian ham via tandem mass spectrometry. *Food Chemistry*, *140*(1), 183–188.

Horák, T., Čulík, J., Čejka, P., Jurková, M., Kellner, V., Dvořák, J. and Hašková, D. (2009). Analysis of Free Fatty Acids in Beer: Comparison of Solid-Phase Extraction, Solid-Phase Microextraction, and Stir Bar Sorptive Extraction. *Journal of Agricultural and Food Chemistry*, *57*(23), 11081–11085. http://doi.org/10.1021/jf9028305.

Horák, T., Čulík, J., Jurková, M., Čejka, P. and Kellner, V. (2008). Determination of free medium-chain fatty acids in beer by stir bar sorptive extraction. *Journal of Chromatography A, 1196-1197*, 96–99. http://doi.org/10.1016/j.chroma.2008.05.014.

Jurado-Sánchez, B., Ballesteros, E. and Gallego, M. (2011). Gas chromatographic determination of 29 organic acids in foodstuffs after continuous solid-phase extraction. *Talanta*, *84*(3), 924–930. http://doi.org/10.1016/j.talanta.2011.02.031.

Käkölä, J., Alén, R., Pakkanen, H., Matilainen, R. and Lahti, K. (2007). Quantitative determination of the main aliphatic carboxylic acids in wood kraft black liquors by high-performance liquid chromatography–mass spectrometry. *Journal of Chromatography A*, *1139*(2), 263–270.

Kaminarides, S., Stamou, P. and Massouras, T. (2007). Comparison of the characteristics of set type yoghurt made from ovine milk of different fat content. *International Journal of Food Science and Technology*, *42*(9), 1019–1028. http://doi.org/10.1111/j.1365-2621.2006.01320.x.

Koohmaraie, M. and Geesink, G. H. (2006). Contribution of postmortem muscle biochemistry to the delivery of consistent meat quality with particular focus on the calpain system. *Meat Science*, *74*(1), 34–43.

Kristoffersen, S., Tobiassen, T., Steinsund, V. and Olsen, R. L. (2006). Slaughter stress, postmortem muscle pH and rigor development in farmed Atlantic cod (Gadus morhua L.). *International Journal of Food Science and Technology*, *41*(7), 861–864.

Kritsunankul, O., Pramote, B. and Jakmunee, J. (2009). Flow injection on-line dialysis coupled to high performance liquid chromatography for the determination of some organic acids in wine. *Talanta*, *79*(4), 1042–1049.

Leite, A. M. O., Leite, D. C. A., Del Aguila, E. M., Alvares, T. S., Peixoto, R. S., Miguel, M. A. L., … Paschoalin, V. M. F. (2013). Microbiological and chemical characteristics of Brazilian kefir during fermentation and storage processes. *Journal of Dairy Science*, *96*(7), 4149–4159. http://doi.org/10.3168/jds.2012-6263.

Li, S., Kundu, D. and Holley, R. A. (2015). Use of lactic acid with electron beam irradiation for control of Escherichia coli O157: H7, non-O157 VTEC E. coli, and Salmonella serovars on fresh and frozen beef. *Food Microbiology*, *46*, 34–39.

Madureira, A. R., Soares, J. C., Amorim, M., Tavares, T., Gomes, A. M., Pintado, M. M. and Malcata, F. X. (2012). Bioactivity of probiotic whey cheese: characterization of the content of peptides and organic acids. *Journal of the Science of Food and Agriculture*, n/a–n/a. http://doi.org/10.1002/jsfa.5915.

Maijala, R. L., Eerola, S. H., Aho, M. A. and Hirn, J. A. (1993). The effect of GDL-induced pH decrease on the formation of biogenic amines in meat. *Journal of Food Protection*, *56*(2), 125–129.

Mani-López, E., García, H. S. and López-Malo, A. (2012). Organic acids as antimicrobials to control Salmonella in meat and poultry products. *Food Research International*, *45*(2), 713–721. http://doi.org/10.1016/j.foodres.2011.04.043.

Mejlholm, O. and Dalgaard, P. (2007). Modeling and Predicting the Growth of Lactic Acid Bacteria in Lightly Preserved Seafood and Their Inhibiting Effect on Listeria monocytogenes. *Journal of Food Protection*, *70*(11), 2485–2497.

Mejlholm, O., Gunvig, A., Borggaard, C., Blom-Hanssen, J., Mellefont, L., Ross, T., … Dalgaard, P. (2010). Predicting growth rates and growth boundary of Listeria monocytogenes — An international validation study with focus on processed and ready-to-eat meat and seafood. *International Journal of Food Microbiology*, *141*(3), 137–150. http://doi.org/10.1016/j.ijfoodmicro.2010.04.026.

Mohan, A. and Pohlman, F. W. (2016). Role of organic acids and peroxyacetic acid as antimicrobial intervention for controlling Escherichia coli O157: H7 on beef trimmings. *LWT-Food Science and Technology*, *65*, 868–873.

Murtaza, M. A., Rehman, S. U., Anjum, F. M., Huma, N., Tarar, O. M. and Mueen-Ud-Din, G. (2012). Organic acid contents of buffalo milk cheddar cheese as influenced by accelerated ripening and sodium salt. *Journal of Food Biochemistry*, *36*(1), 99–106. http://doi.org/10.1111/j.1745-4514.2010.00517.x.

Ong, L., Henriksson, A. and Shah, N. P. (2006). Development of probiotic Cheddar cheese containing Lactobacillus acidophilus, Lb. casei, Lb. paracasei and Bifidobacterium spp. and the influence of these bacteria on proteolytic patterns and production of organic acid. *International Dairy Journal*, *16*(5), 446–456. http://doi.org/10.1016/j.idairyj.2005.05.008.

Osako, K., Hossain, M. A., Kuwahara, K., Okamoto, A., Yamaguchi, A. and Nozaki, Y. (2005). Quality aspect of fish sauce prepared from underutilized fatty Japanese anchovy and rabbit fish. *Fisheries Science*, *71*(6), 1347–1355.

Quirós, A. R.-B., Lage-Yusty, M. A. and López-Hernández, J. (2009). HPLC analysis of organic acids using a novel stationary phase. *Talanta*, *78*(2), 643–646. http://doi.org/10.1016/j.talanta.2008.11.013.

Quitmann, H., Fan, R. and Czermak, P. (2013). Acidic Organic Compounds in Beverage, Food, and Feed Production. In H. Zorn and P. Czermak (Eds.),

Biotechnology of Food and Feed Additives (pp. 91–141). Springer Berlin Heidelberg. Retrieved from http://link.springer.com/chapter/10.1007/10_2013_262.

Ricke, S. C. (2003). Perspectives on the use of organic acids and short chain fatty acids as antimicrobials. *Poultry Science*, *82*(4), 632–639.

Saithong, P., Panthavee, W., Boonyaratanakornkit, M. and Sikkhamondhol, C. (2010). Use of a starter culture of lactic acid bacteria in plaa-som, a Thai fermented fish. *Journal of Bioscience and Bioengineering*, *110*(5), 553–557. http://doi.org/10.1016/j.jbiosc.2010.06.004.

Saraji, M. and Mirmahdieh, S. (2009). Single-drop microextraction followed by in-syringe derivatization and GC-MS detection for the determination of parabens in water and cosmetic products. *Journal of Separation Science*, *32*(7), 988–995.

Saraji, M. and Mousavinia, F. (2006). Single-drop microextraction followed by in-syringe derivatization and gas chromatography-mass spectrometric detection for determination of organic acids in fruits and fruit juices. *Journal of Separation Science*, *29*(9), 1223–1229.

Sriphochanart, W. and Skolpap, W. (2011). The use of selected lactic acid bacteria starter cultures for improved Thai Sausage Fermentation. *Journal of Food Processing and Preservation*, *35*(3), 291–298. http://doi.org/10.1111/j.1745-4549.2009.00453.x.

Suárez-Luque, S., Mato, I., Huidobro, J. F. and Simal-Lozano, J. (2002). Solid-phase extraction procedure to remove organic acids from honey. *Journal of Chromatography B*, *770*(1), 77–82.

Subramanian, A., Alvarez, V. B., Harper, W. J. and Rodriguez-Saona, L. E. (2011). Monitoring amino acids, organic acids, and ripening changes in Cheddar cheese using Fourier-transform infrared spectroscopy. *International Dairy Journal*, *21*(6), 434–440.

Suomalainen, T. H. and Mäyrä-Makinen, A. M. (1999). Propionic acid bacteria as protective cultures in fermented milks and breads. *Le Lait*, *79*(1), 165–174.

Swartz, M. (2010). HPLC detectors: a brief review. *Journal of Liquid Chromatography and Related Technologies*, *33*(9-12), 1130–1150. http://doi.org/10.1080/10826076.2010.484356.

Swetwiwathana, A. and Visessanguan, W. (2015). Potential of bacteriocin-producing lactic acid bacteria for safety improvements of traditional Thai fermented meat and human health. *Meat Science*, *109*, 101–105.

Theron, M. M. and Lues, J. R. (2010). *Organic acids and food preservation*. CRC

Tormo, M. and Izco, J., (2004). Alternative reversed-phase high-performance liquid chromatography method to analyse organic acids in dairy products. *Journal of Chromatography A*, *1033*(2), 305–310. http://doi.org/10.1016/j.chroma.2004.01.043.

van Hees, P. A., Dahlén, J., Lundström, U. S., Borén, H. and Allard, B. (1999). Determination of low molecular weight organic acids in soil solution by HPLC. *Talanta*, *48*(1), 173–179.

Walstra, P. (2013). *Dairy technology: principles of milk properties and processes*. CRC Press.

Wen, Y., Wang, Y. and Feng, Y.-Q. (2007). Extraction of clenbuterol from urine using hydroxylated poly (glycidyl methacrylate-co-ethylene dimethacrylate) monolith microextraction followed by high-performance liquid chromatography determination. *Journal of Separation Science*, *30*(17), 2874–2880.

Yang, J. and Caprioli, R. M. (2011). Matrix sublimation/recrystallization for imaging proteins by mass spectrometry at high spatial resolution. *Analytical Chemistry*, *83*(14), 5728–5734.

Yang, M.-H. and Choong, Y.-M. (2001). A rapid gas chromatographic method for direct determination of short-chain (C< sub> 2</sub>–C< sub> 12</sub>) volatile organic acids in foods. *Food Chemistry*, *75*(1), 101–108.

Zaki, H. M., Mohamed, H. M. and El-Sherif, A. M. (2015). Improving the antimicrobial efficacy of organic acids against Salmonella enterica attached to chicken skin using SDS with acceptable sensory quality. *LWT-Food Science and Technology*, *64*(2), 558–564.

Zeppa, G., Conterno, L. and Gerbi, V. (2001). Determination of Organic Acids, Sugars, Diacetyl, and Acetoin in Cheese by High-Performance Liquid Chromatography. *Journal of Agricultural and Food Chemistry*, *49*(6), 2722–2726. http://doi.org/10.1021/jf0009403.

Biographical Sketches

Marion Pereira da Costa

Affiliation: Universidade Federal Fluminense and Universidade Federal do Rio de Janeiro

Education: Degree of Medicine Veterinarian (D.V.M.); Master of Science (M.Sc.); Philosophic Doctor (Ph.D.).

Research and Professional Experience: Professor of Food Science in the Faculty of Nutrition of Universidade Federal do Rio de Janeiro

Professional Appointments: Researcher of Food Technology of Faculty of Veterinarian of Universidade Federal Fluminense and Post-doctored of Food Science of Chemistry Institute of Universidade Federal do Rio de Janeiro

Publications Last 3 Years:

Costa, MP; Frasao, BS; Costa-Lima, BRC; Rodrigues, BLL; Conte-Junior, CA. Simultaneous analysis of carbohydrates and organic acids by HPLC-DAD-RI for monitoring goat's milk yogurts fermentation. Talanta (Oxford), 152, 162-170, 2016.

Balthazar, CF; Conte-Junior, CA; Moraes, J; Costa, MP.; Raices, RSL; Franco, RM; Cruz, AG; Silva, ACO. Physicochemical evaluation of sheep milk yogurts containing different levels of inulin. Journal of Dairy Science, 99, 4160-4168, 2016.

Cutrim, CS; Barros, RF; Costa, MP; Franco, RM; Conte-Junior, CA; Cortez, MAS. Survival of Escherichia coli O157:H7 during manufacture and storage of traditional and low lactose yogurt. Lebensmittel-Wissenschaft + Technologie / Food Science + Technology, 70, 178-184, 2016.

Machado, STZ; Rezende, AR; Gennari, S; Conte-Junior, CA; Costa, MP; Lázaro, CAT; Telles, EO. Development of HPLC-Fluorescence Method for the Determination of Ivermectin Residues in Commercial Milk. Journal of Experimental Food Chemistry, 2, 2016.

Costa, MP; Conte-Junior, CA. Chromatographic Methods for the Determination of Carbohydrates and Organic Acids in Foods of Animal Origin. Comprehensive Reviews in Food Science and Food Safety, 14, 586-600, 2015.

Costa, MP; Frasao, BS; Silva, ACO; Freitas, MQ; Franco, RM; Conte-Junior, CA. Cupuassu (*Theobroma grandiflorum*) pulp, probiotic, and prebiotic: Influence on color, apparent viscosity, and texture of goat milk yogurts. Journal of Dairy Science, 98, 5995-6003, 2015.

Costa, MP; Balthazar, CF; Rodrigues, BL; Lazaro, CA; Silva, ACO; Cruz, AG; Conte-Junior, CA. Determination of biogenic amines by high-performance liquid chromatography (HPLC-DAD) in probiotic cow's and goat's fermented milks and acceptance. Food Science and Nutrition, 3, 1-7, 2015.

Gaze, LV; Costa, MP; Monteiro, MLG; Lavorato, JAA; Conte-Junior, CA; Raices, RSL; Cruz, AG; Freitas, MQ. Dulce de Leche, a typical product of Latin America: Characterisation by physicochemical, optical and instrumental methods. Food Chemistry, 169, 471-477, 2015.

Silva, HLA; Costa, MP; Frasao, BS; Mesquita, EFM; Mello, SCRP; Conte-Junior, CA; Franco, RM; Miranda, ZB. Efficacy of Ultraviolet-C Light to Eliminate S taphylococcus Aureus on Precooked Shredded Bullfrog Back Meat. Journal of Food Safety, v. 35, 1-6, 2015.

Almeida, CC; Alvares, TS; Costa, MP; Conte-Junior, CA. Protein and Amino Acid Profiles of Different Whey Protein Supplements. Journal of Dietary Supplements, 13, 313-323, 2015.

Costa, MP; Balthazar, CF; Franco, RM; Mársico, ET; Cruz, AG; Conte-Junior, CA. Changes on expected taste perception of probiotic and conventional yogurts made from goat milk after rapidly repeated exposure. Journal of Dairy Science, v. 97, p. 2610-2618, 2014.

Costa, MP; Balthazar, CF; Moreira, RVBP; Cruz, AG; Conte-Junior, CA. Leite fermentado: potencial alimento funcional. Enciclopédia Biosfera, 9, 1387-1408, 2013.

Rodrigues, BL; Alvares, TS; Costa, MP; Sampaio, GSL; Lazaro, CA; Marsico, ET; Conte-Junior, CA. Concentration of Biogenic Amines in Rainbow Trout (Oncorhynchus mykiss) preserved in ice and its Relationship with Pysicochemical Parameters of Quality. Journal of Aquaculture Research and Development, 4, 1-4, 2013.

Costa, MP; Silva, HLA; Balthazar, CF; Franco, RM; Cortez, MAS. Economic performance and sensory analysis of probiotic Minas Frescal Cheese produced using bovine and caprine milk. Enciclopédia Biosfera, 9, 2306-2314, 2013.

Costa, MP; Conte-Junior, CA. Leites fermentados como alimentos funcionais. Animal Business Brasil, 3, 60, 2013.

Carlos Adam Conte-Junior

Affiliation: Universidade Federal Fluminense and Universidade Federal do Rio de Janeiro

Education: Degree of Medicine Veterinarian (D.V.M.) of Universidade Federal Fluminense; Master of Science (M.Sc.) in Food Science in the

Universidade Federal do Rio de Janeiro; Philosophy Doctor (Ph.D.) in Medicine Veterinary in the Universidade Federal Fluminense and Nutrition and Food Science by the Universidad Complutense de Madrid (Madrid, Spain).

Research and Professional Experience:

2014-2015	Post-doctorate of Gemone Center, University of California, Davis.
2014-2015	Visiting professor of Department of Food Science and Technology, University of California, Davis.
2008-2011	Researcher, Department of Biochemistry, Rio de Janeiro Federal University, Brazil.
2008	Visiting Scientist (7 months) in Karolinska Institutet, working with Characterization/identification of novel fatty acids secreted by probiotic as inducers of FIAF expression in HT-29 cells. Stockholm, Sweden.
2007	Visiting Scientist (5 months) in Medish Centrum at the Vrije Universiteit Amsterdam working with analysis of folic acid produced by probiotic bacteria. Amsterdam, Netherlands.
2007	Visiting Scientist (2 months) in Dipartimento di Scienze e Tecnologie Veterinarie per la Sicurezza Alimentar at the Universita Degli Studi di Milano working with chromatographic techniques in food analysis. Milan, Italy.

Professional Appointments:

Professor of Department of Food Technology, Universidade Federal Fluminense (UFF)

Professor of Food Science Program, Universidade Federal do Rio de Janeiro (UFRJ)

Scientist of Research Foundation of the State of Rio de Janeiro (FAPERJ)

Researcher of National Council of Technological and Scientific Development (CNPq)

Honors:

2014 Scientist Award, Rio de Janeiro Research Supporting Foundation.

2013 Researcher Award of National Council of Technological and Scientific Development.

2012 Young Scientist Award, Rio de Janeiro Research Supporting Foundation.

2011 First Place Award, Oral Presentation, 21th Seminary Vasconcelos Torres, awarded by the National Council for Scientific and Technological Development.

2011 Outstanding Paper Presentation Award, 34th Annual Meeting of Brazilian Chemical Society, awarded by the Brazilian Chemical Society.

2010 Outstanding Paper Presentation Award, 7th Brazilian Congress of Food Microbiology, awarded by the Brazilian Society of Microbiology.

Publications Last 3 Years:

Costa, MP; Frasao, BS; Costa-Lima, BRC; Rodrigues, BLL; Conte-Junior, CA. Simultaneous analysis of carbohydrates and organic acids by HPLC-DAD-RI for monitoring goat's milk yogurts fermentation. Talanta (Oxford), 152, 162-170, 2016.

Balthazar, CF; Conte-Junior, CA; Moraes, J; Costa, MP; Raices, RSL; Franco, RM; Cruz, AG; Silva, ACO. Physicochemical evaluation of sheep milk yogurts containing different levels of inulin. Journal of Dairy Science, 99, 4160-4168, 2016.

Cutrim, CS; Barros, RF; Costa, MP; Franco, RM; Conte-Junior, CA; Cortez, MAS. Survival of Escherichia coli O157:H7 during manufacture and storage of traditional and low lactose yogurt. Lebensmittel-Wissenschaft + Technologie / Food Science + Technology, 70, 178-184, 2016.

Machado, STZ; Rezende, AR; Gennari, S; Conte-Junior, CA; Costa, MP; Lázaro, CAT; Telles, EO. Development of HPLC-Fluorescence Method for the Determination of Ivermectin Residues in Commercial Milk. Journal of Experimental Food Chemistry, 2, 2016.

Rodrigues, BL; Alvares, TS; Sampaio, GSL; C, CC; Araujo, JVA; Franco, RM; Mano, SB; Conte-Junior, CA. Influence of vacuum and modified atmosphere packaging in combination with UV-C radiation on the shelf life of rainbow trout (Oncorhynchus mykiss) fillets. Food Control, 60, 596-605, 2016.

Guedes-Oliveira, JM; Salgado, RL; Costa-Lima, BRC; Guedes-Oliveira, J; Conte-Junior, CA. Washed cashew apple fiber (Anacardium occidentale L.) as fat replacer in chicken patties. Lebensmittel-Wissenschaft + Technologie / Food Science + Technology, 71, 268-273, 2016.

Canto, ACVCS; Costa-Lima, BRC; Suman, SP; Montieiro, MLG; Viana, FM; Salim, APAA; Nair, MN; Silva, TJP; Conte-Junior, CA. Color attributes and oxidative stability of longissimus lumborum and psoas major muscles from Nellore bulls. Meat Science, 121, 19-26, 2016.

Carneiro, CS; Mársico, ET; Ribeiro, ROR; Conte-Júnior, CA; Mano, SB; Augusto, CJC; Oliveira de Jesus, EF. Low-Field Nuclear Magnetic Resonance (LF NMR 1H) to assess the mobility of water during storage of salted fish (Sardinella brasiliensis). Journal of Food Engineering, 169, 321-325, 2016.

Leonardo, R; Nunes, RSC; Montieiro, MLG; Conte-Junior, CA; Del Aguila, EM; Paschoalin, VMF. Molecular testing on sardines and rulings on the authenticity and nutritional value of marketed fishes: An experience report in the state of Rio de Janeiro, Brazil. Food Control, 60, 394-400, 2016.

Felicio, TL; Esmerino, EA; Vidal, VAS; Cappato, LP; Garcia, RKA; Cavalcanti, RN; Freitas, MQ; Conte-Junior, CA; Padilha, MC; Silva, MC; Raices, RSL; Arellano, DB; Bollini, HMA; Pollonio, MAR; Cruz, AG. Physico-chemical changes during storage and sensory acceptance of low sodium probiotic Minas cheese added with arginine. Food Chemistry, 196, 628-637, 2016.

da Silva, DVT; Silva, FO; Perrone, D; Pierucci, APTR; Conte-Junior, CA; Alvares, TS; Aguila, EMD; Paschoalin, VMF. Physicochemical, nutritional, and sensory analyses of a nitrate-enriched beetroot gel and its effects on plasmatic nitric oxide and blood pressure. Food and Nutrition Research, 60, 29909, 2016.

Costa, MP; Conte-Junior, CA. Chromatographic Methods for the Determination of Carbohydrates and Organic Acids in Foods of Animal Origin. Comprehensive Reviews in Food Science and Food Safety, 14, 586-600, 2015.

Costa, MP; Frasao, BS; Silva, ACO; Freitas, MQ; Franco, RM; Conte-Junior, CA. Cupuassu (*Theobroma grandiflorum*) pulp, probiotic, and prebiotic: Influence on color, apparent viscosity, and texture of goat milk yogurts. Journal of Dairy Science, 98, 5995-6003, 2015.

Costa, MP; Balthazar, CF; Rodrigues, BL; Lazaro, CA; Silva, ACO; Cruz, AG; Conte-Junior, CA. Determination of biogenic amines by high-performance liquid chromatography (HPLC-DAD) in probiotic cow's and goat's fermented milks and acceptance. Food Science and Nutrition, 3, 1-7, 2015.

Gaze, LV; Costa, MP; Montieiro, MLG; Lavorato, JAA; Conte-Junior, CA; Raices, RSL; Cruz, AG; Freitas, MQ. Dulce de Leche, a typical product of Latin America: Characterisation by physicochemical, optical and instrumental methods. Food Chemistry, 169, 471-477, 2015.

Silva, HLA; Costa, MP; Frasao, BS; Mesquita, EFM; Mello, SCRP; Conte-Junior, CA; Franco, RM; Miranda, ZB. Efficacy of Ultraviolet-C Light to Eliminate S taphylococcus Aureus on Precooked Shredded Bullfrog Back Meat. Journal of Food Safety, v. 35, 1-6, 2015.

Almeida, CC; Alvares, TS; Costa, MP; Conte-Junior, CA. Protein and Amino Acid Profiles of Different Whey Protein Supplements. Journal of Dietary Supplements, 13, 313-323, 2015.

Vieira, CP; Álvares, TS; Gomes, LS; Torres, AG; Paschoalin, VMF; Conte-Junior, CA. Kefir Grains Change Fatty Acid Profile of Milk during Fermentation and Storage. Plos One, 10, 2015.

Macedo, F; Mársico, ET; Conte-Júnior, CA; de Resende, MF; Brasil, TF; Netto, ADP. Development and validation of a method for the determination of low-ppb levels of macrocyclic lactones in butter, using HPLC-fluorescence. Food Chemistry, 179, 239-245, 2015.

Montieiro, MLG; Mársico, ET; Lázaro, CA; Canto, ACVC; Lima, BRCC; Cruz, AG; Conte-Júnior, CA. Effect of transglutaminase on quality characteristics of a value-added product tilapia wastes. Journal of Food Science and Technology, 52, 2598-2609, 2015.

Macedo, F; Mársico, ET; Conte-Júnior, CA; Furtado, LA; Brasil, TF; Netto, ADP. Short communication: Macrocyclic lactone residues in butter from Brazilian markets. Journal of Dairy Science, 98, 3695-3700, 2015.

Costa, MP; Balthazar, CF; Franco, RM; Mársico, ET; Cruz, AG; Conte-Junior, CA. Changes on expected taste perception of probiotic and conventional yogurts made from goat milk after rapidly repeated exposure. Journal of Dairy Science, v. 97, p. 2610-2618, 2014.

Ribeiro, ROR; Mársico, ET; Jesus, EFO; Carneiro, CS; Conte-Junior, CA; Almeida, E; Nascimento Filho, VF. Determination of Trace Elements in Honey from Different Regions in Rio de Janeiro State (Brazil) by Total Reflection X-Ray Fluorescence. Journal of Food Science, 79, T738-T742, 2014.

Alvares, TS; Conte-Junior, CA; Silva, JT; Paschoalin, VMF. l-arginine does not improve biochemical and hormonal response in trained runners after 4 weeks of supplementation. Nutrition Research (New York, N.Y.), 34, 31-39, 2014.

Mársico, ET; Ferreira, MS; São Clemente, SC; Gouvea, RCS; Jesus, EFO; Conti, CC; Conte-Junior, CA; Kelecom, AGAC. Distribution of Po-210 in two species of predatory marine fish from the Brazilian coast. Journal of Environmental Radioactivity, 128, 91-96, 2014.

Ribeiro, ROR; Mársico, ET; Carneiro, CS; Montieiro, MLG; Conte-Júnior, CA; Jesus, EFO. Detection of honey adulteration of High Fructose Corn Syrup by Low field Nuclear Magnetic Resonance (LF 1H NMR). Journal of Food Engineering, 135, 39-43, 2014.

Costa, MP; Balthazar, CF; Moreira, RVBP; Cruz, AG; Conte-Junior, C.A. Leite fermentado: potencial alimento funcional. Enciclopédia Biosfera, 9, 1387-1408, 2013.

Rodrigues, BL; Alvares, TS; Costa, MP; Sampaio, GSL; Lazaro, CA; Mársico, ET; Conte-Junior, CA. Concentration of Biogenic Amines in Rainbow Trout (Oncorhynchus mykiss) preserved in ice and its Relationship with Pysicochemical Parameters of Quality. Journal of Aquaculture Research and Development, 4, 1-4, 2013.

Costa, MP; Conte-Junior, CA. Leites fermentados como alimentos funcionais. Animal Business Brasil, 3, 60, 2013.

Lázaro, CALT; Conte-Junior, CA; Cunha, FL; Mársico, ET; Mano, SB; Franco, RM. Validation of an HPLC methodology for the identification and quantification of biogenic amines in chicken meat. Food Analytical Methods (Print), 6, 1024-1032, 2013.

In: Organic Acids
Editor: Cesar Vargas
ISBN: 978-1-63485-931-8

Chapter 3

Diversity of Organic Acids Content in Green and Roasted *Coffea arabica* Cultivars

***Cíntia S. G. Kitzberger*[1,*], *Maria Brígida S. Scholz*[1], *João BGD Silva*[2] *and Marta T Benassi*[3]**

[1]Área de Ecofisiologia, Laboratório de Fisiologia Vegetal, Instituto Agronômico do Paraná, Londrina, Paraná, Brazil

[2] Centro Tecnológico Cocari, Mandaguari, Paraná, Brazil

[3]Departamento de Ciência e Tecnologia de Alimentos, Universidade Estadual de Londrina, Londrina, Paraná, Brazil

Abstract

Organic acids in coffees are influenced by several factors. The aim of this research was to analyze the profile of organic acids (quinic, malic, citric, acetic, and lactic) and chlorogenic acids in green and roasted coffees cultivar and pH and titrable acidity (TA) in brew coffee. Sixteen cultivars (traditional and modern) grown and harvested in the same place were evaluated. Hierarchical cluster analysis of green coffees formed three groups. G1-Bourbon, Catuaí, Icatu and Catuaí SH2SH3 derived cultivars were associated to higher malic, citric and intermediate value of quinic and 5-CQA. G2 (Catuaí, Icatu) containing higher level of quinic,

* Corresponding author: Cíntia S. G. Kitzberger. E-mail: cintiasorane@yahoo.com.br.

citric acid and lower value of malic and 5-CQA and G3 (Sarchimor derived) presented lower values of quinic, citric and higher content of 5-CQA. Roasted coffee and brews characteristics formed three groups. G1 (IPR 99, Icatu and Catuaí SH2SH3 derived) with higher values of quinic, pH and intermediate acetic acid values and lower malic, citric, 5-CQA, lactic, and TA. G2 (Bourbon, IPR 97) presented higher content of malic, latic, 5-CQA acids and acidity and intermediate of citric and lower of quinic, acetic acids and pH. G3 (Catuaí, Icatu, Sarchimor, Catuaí and Icatu derived) showed higher value of quinic, acetic, citric, and intermediate values of 5-CQA, lactic, malic acids, pH and TA. We can verify that the TA did not differ between the groups but different acids contributed for the formation of TA. It was verified that roasting process provides different acids profiles associated to genetic origin of cultivars.

Keywords: malic acid, citric acid, quinic acid, coffee brews acidity, genetic background

Introduction

Organic acids content, mainly the free form, contribute to the acidity of the coffee brews and therefore for their sensory quality. Genetic background (coffee species and cultivars), conditions of growing, stage of maturation, post-harvesting, roasting and storage conditions, and brewing processes can influence the organic acids profile in coffee brews [1, 2]. The acids are responsible for acidity, which together with aroma and bitter are the main coffee brews attributes [3, 4]. Some acids like citric, malic and quinic are naturally present in the green coffee beans, other such as acetic and lactic acids were generated during roasting process from carbohydrate precursors, mainly sucrose [5, 6]. Chlorogenic acids can be found in different quantities depending on the degree of maturation of coffee beans and they were associated to the brew quality [7, 8]. Quinic acid can be found in green coffee and is also formed during the roasting [2].

Organic acids play an important role in plant development, assisting the chelation and neutralization of the toxicity of aluminum, promoting a rapid adaptation of cellular metabolism and also activating and attaching potential nutrients around the plant roots [9].

Breeding programs usually focus their efforts on the transference of genes from *Coffea canephora* to *C. arabica* in order to increase the resistance of arabica cultivars against pests and diseases. However, these crosses can also modify the composition of coffees affecting, for example, the organic acids profile. Thus, it is important to evaluate the effect of the genetic background on the composition for new coffee crosses.

The Instituto Agronômico of Paraná (IAPAR) has developed different arabica cultivars derived from crosses of *C. arabica* Villa Sarchi × Timor Hybrid (Sarchimor): Iapar 59, IPR 98 and IPR 99) which are resistant to rust and IPR 100 and IPR 106, which were resistant to *Meloidogyne paranaensis*. Other crossings between Icatu and Catuaí resulted in a cultivar (IPR 102) resistant to bacterial blight disease or bacteriosis were also developed [10, 11].

The objective of the research was to compare the organic acids and 5-CQA profile of thirteen new crosses (IAPAR 59, and IPRs 97, 98, 99, 100, 101, 102, 103, 104, 105, 106, 107, and 108) to traditional ones (Red Bourbon, Red Catuaí, and Yellow Icatu). In order to achieve a broader view of the matter, green coffee beans and roasted coffees were evaluated. Some characteristics of the coffee brews associated with the organic acids profile (titratable acidity and pH) were also reported.

Table 1. Cultivars and their genetic background

Cultivars		Genetic background
Traditional	Red Bourbon	Pure arabica
	Red Catuaí	Yellow Caturra (simple mutation of Red Bourbon) x Mundo Novo (hybridization between Red Bourbon and Sumatra)
	Yellow Icatu	Red Icatu (hybrid of robusta and arabica) x Mundo Novo x Yellow Bourbon
Modern crosses	Iapar 59, IPR 97, 98, 99, 104	Timor Hybrid and Villa Sarchi (Sarchimor)
	IPR 100, 101, 105	Derived from a cross of Catuaí Sh_2 Sh_3
	IPR 102	Icatu x Catuaí
	IPR 103	Red Catuaí IAC 99 and Yellow IAC 66 x Icatu
	IPR 106	Icatu
	IPR 107	Sarchimor x Mundo Novo
	IPR 108	Sarchimor x Icatu x Catuaí

Reference: [10-16].

Experimental Section

Coffees of different genetic background were studied (Table 1). Coffees were harvested at the Agricultural Technologic Park of Cooperative COCARI, Mandaguari, Paraná, Brazil, from May to July 2009. Coffees were grown at latitude (S) 23°32'52", altitude of 650 m and average annual temperatures of 22 to 23°C. Harvesting and post-harvesting conditions were standardized for all cultivars. The time of harvesting was variable according to the maturation stage for each cultivar. Coffees collected at the cherry stage were naturally sun-dried. After hulling (removing husks and parchment), the green coffee beans were standardized using a size 16 sieve (diameter 6.5 mm) and defective beans were removed.

A part of the samples of green coffee beans were frozen using liquid nitrogen to prevent oxidation of compounds and immediately were grounded (0.5 mm particles) in the disk mill (PERTEN 3600, Sweden). The milled samples were stored in -18°C, until chemical determination.

Another part of the green coffee beans were subjected to medium roasting process (8 to 11 minutes at 200-210°C) in roaster for small samples (Rod-Bel roaster, Brazil). The degree of roasting was controlled by weight loss and lightness (L *) of roasted and ground coffee. The roasting losses ranged from 13 to 14% and lightness (L *) was around 28.

The roasted coffee was ground (with particles size not exceeding 0.6 mm) and frozen until analysis. For pH and titratable analyses, the brews were prepared with 70 g of roasted coffee L^{-1} and it was filtered through filter paper.

Analytical standard-grade organic acids were used: quinic, malic, citric, latic and acetic acids (Sigma Aldrich, St. Louis, MI, USA). Stock solutions were prepared by dissolution of acids in ultra-pure water.

Samples were filtered through HN nylon membrane syringe filters (13 mm, 0.45 μm) (Milipore, Billerica, MA, USA) and cartridge SPE was assembled with Dowex 1X4 200 mesh ion-exchange resin (Sigma Aldrich, St. Louis, MI, USA) and extraction was performed in a Vac-Elut Varian Manifold. HPLC mobile phase was prepared with sulfuric acid 0,005N (Vetec, Rio de Janeiro, RJ, Brazil). HPLC analyses were carried out in a Surveyor Plus liquid chromatograph (Thermo Scientific, San Jose, USA) consisting of a Peltier autosampler with temperature control and an integrated oven (Surveyor Plus), a quaternary pump (Surveyor LC Plus) and a diode array detector (Surveyor PDA Plus). The equipment was coupled to an interface (SS420) and a ChromQuest 5.0 chromatography data system.

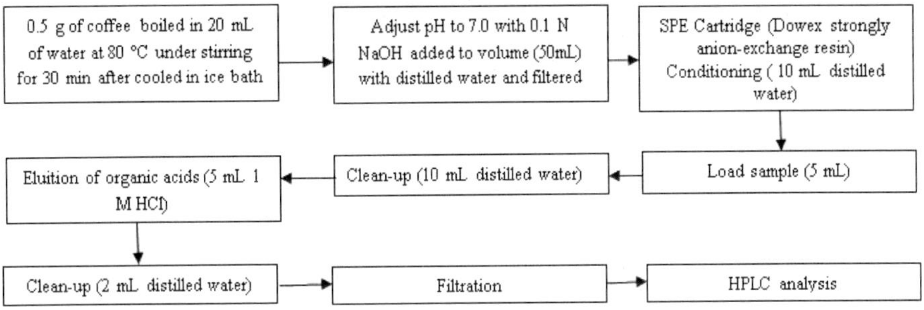

Figure 1. Flow-chat of organic acids extraction.

Table 2. Chromatographic parameters, linearity, sensitivity and precision of the method for organic acids

	RT (min)	Linear regression[a]			Sensitivity		Recovery (%)[b]
		Range (mg 100 g^{-1})	Equation	R^2	LOD (mg 100 g^{-1})	LOQ (mg 100 g^{-1})	
Quinic	5.54	125-2000	y = 403.2 x + 43448	0.997	0.04	0.13	84
Malic	6.05	50-800	y =629 x - 18369	0.98	0.10	0.30	78
Citric	8.66	125-2000	y = 916.5 x + 10350	0.995	0.03	0.10	88
Latic	6.86	25-400	y = 280.8 x -935.4	0.998	0.08	0.23	79
Acetic	7.16	20-300	y = 348 x -1065	0.996	0.01	0.02	60

[a] y = concentration in mg/100 g sample; x = peak area; n = 3.

[b] Analysis in duplicate; standards were added in an amount of approximately 50% of the initial content.

Quinic, malic, lactic, acetic and citric acids were extracted and quantified as described by Kitzberger et al. [17]. Analysis details were shown in Figure 1.

The HPLC analysis was performed using an ACE 5 C18 column (250 mm x 4.6 mm id, 5 mm) (Advanced Chromatography Technologies, Aberdeen) with detection at 210 nm. Isocratic elution of 0.005 N H_2SO_4 solution (pH 2.5) was carried out with a gradient of flow rate: 0.7 mL min^{-1} from 0 to2 min; 0.4 mL min^{-1} from 2 to 15 min; and 0.7 mL min^{-1}up to 15 min. Oven temperature of 30°C and temperature of the sample tray of 5°C were applied. Green and roasted coffees were evaluated and identification of acids was done by comparison with standards and spiking. Some chromatographic parameters, linearity, sensitivity and precision of the method for each organic acid are shown in Table 2.

Chlorogenic acid (5-CQA) was quantified, by HPLC, in green and roasted coffees, as described by Alves et al. [18]. pH and titratable acidity were only measured in the brew coffee. pH was determined in 10 mL of coffee brew after reaching 25°C in a digital pH meter (Metrohm, model 744). Titrable acidity was determined in 10 mL of coffee brew titrated with 0.1 N NaOH to pH 8.2. The result was expressed in 0.1 mL of NaOH to 100 mL of brew.

Principal Components Analysis (PCA) and Hierarchical Clustering Analysis (HCA) were applied to analyze the data, using the XLStat software [19].

Results and Discussion

In a general way, the contents of the acids in green coffees beans varied in the ranges as follows: from 0.35 to 0.55 g of quinic acid 100 g^{-1}; from 0.30 to 0.64 g of malic acid 100 g^{-1}; from 0.93 to 1.31 g of citric acid 100 g^{-1}; and from 4.17 to 5.35 g of 5-CQA 100 g^{-1}. These values were comparable to the reported by Steiman [20] for coffee beans from different origins and crosses: 0.57 g 100 g^{-1} for quinic acid, 0.41 g 100 g^{-1} for malic acid, 1.37 g 100 g^{-1} for citric acid and 3.21 to 6.97 g 100 g^{-1} for 5-CQA.

The organic acids content in coffee beans is highly associated with the stage of maturation [2]. Concentration of quinic and malic acids decrease as the maturation has taken place [2, 21]. High contents of 5-CQA were also usually related to immature beans [22-24]. Citric acid has an opposite behavior, presenting lower values in the initial stage of development of beans with an increase during the maturation [2].

Hierarchical Clustering Analysis (HCA) allows analyzing the formation of groups of cultivars due different composition to green coffee. Considering acid composition of green coffee beans HCA was applied in green coffees cultivars and the result were presented in Figure 2. Three groups were formed. The first group (G1) was formed by Bourbon, IPR 100, IPR 101, IPR 102, IPR 105, IPR 106 and IPR 107 cultivars which were associated to higher malic, citric and intermediate value of quinic and 5-CQA. The second group (G2) was formed by Catuaí and Icatu containing higher level of quinic, citric acid and lower value of malic and 5-CQA. The third one (G3) was formed by Iapar 59, IPR 97, IPR 98, IPR 99, IPR 103 and IPR 108 cultivars and these cultivars presented lower values of quinic, citric and higher content of 5-CQA (Table 3).

Citric acid and 5-CQA were found in similar amounts in the three groups, suggesting that the cultivars of the three groups present similar maturity, since these compounds were associated to maturation.

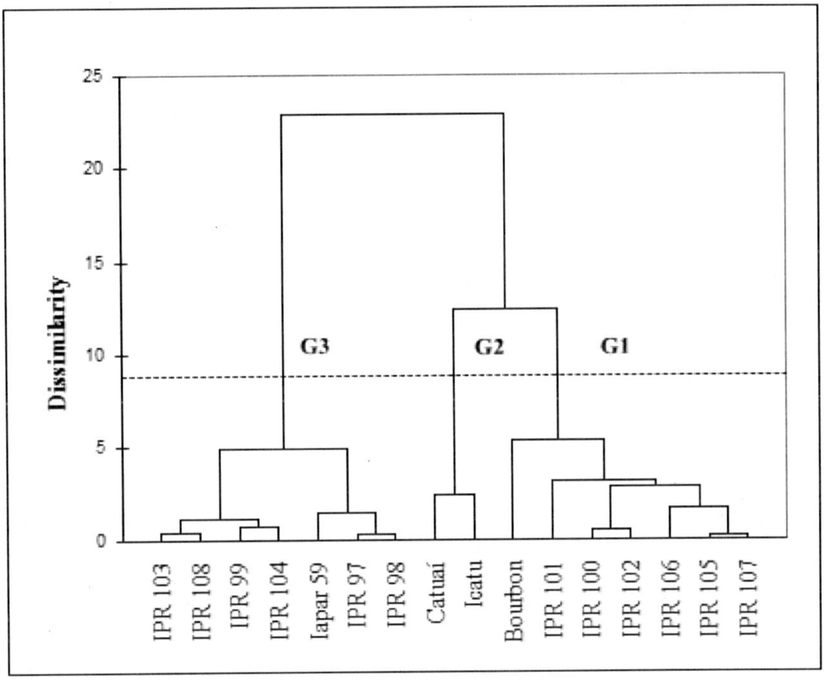

Figure 2. Dendogram of the coffee cultivars considering organic acids composition of green beans.

Table 3. Average value of compounds content for each group formed by HCA for green coffees (g 100 g^{-1})*

	Quinic	Malic	Citric	5-ACQ
G1	0.45± 0.037a	0.56 ± 0.057a	1.17 ± 0.076a	4.59 ± 0.265a
G2	0.50 ± 0.047a	0.31 ± 0.013b	1.17 ± 0.051a	4.35 ± 0.180a
G3	0.37 ± 0.017b	0.50 ± 0.055a	1.02 ± 0.061a	4.91 ±0.229a

* Different small letters in the same column indicate significant differences among groups ($p > 0.05$).

In the Table 3 it was possible to verified that most cultivars with Sarchimor crosses (Table 1) were allocated in G3, and derived cultivars Catuaí and Icatu and Catuaí SH2SH3 were allocated in the G1.

The acid profile in green coffee does not always indicate the quality of roasted coffee. During roasting, organic acids are degraded and others are formed from various reactions. The concentration in organic acids of roasted coffee is highly dependent on the degree of roasting and usually coffee with dark roasting are less acids [25-27]. For comparative studies is therefore necessary to know the degree of roasting.

The cultivars roasted showed a wide range of concentration of acids. In present study, the contents of the acids in roasted coffees varied in the ranges as follows: from 0.66 to 1.05 g of quinic acid 100 g^{-1} of coffee; from 0.05 to 0.35 g of malic acid 100 g^{-1}; from 0.16 to 0.44 g of lactic acid 100 g^{-1}; from 0.14 to 0.32 g of acetic acid 100 g^{-1}; 0.52 to 0.79 g of citric acid 100 g^{-1}; and from 0.82 to 1.93 g of 5-CQA 100 g^{-1}.

Despite the very close values of quinic acid in green coffee (Table 3), quinic acid content in roasted coffee was quite variable. Quinic acid showed an increase during the roasting resulted of mainly degradation of chlorogenic acids [28]. Degradation of quinic acid occurs only in very intense roasting [29].

In the present study the increase in quinic acid were according each cultivar (Figure 3). Quinic acid increased on average by 120%. The largest increase was found in the cultivar Iapar 59 (188%), and the lowest value was observed in the cultivar Bourbon (48%). It was noted great variability in the formation and degradation of acids among the evaluated cultivars and could not find correlation between the degradation of 5-CQA and the formation of quinic acid (Table 4). Studies showed different levels of degradation of 5-CQA cultivars of different genetic origin [2].

The degradation of 5-CQA among cultivars showed substantial variation, ranging between 62% (IPR 97) to 82% (IPR 106). The remaining amount of 5-CQA after roasting was lower for Catuaí, IPR 99, IPR 106 and Catuaí derived cultivars (Ca + SH2SH3) compared to the cultivars derived from Sarchimor (Figure 3). As the coffees were roasted in a similar way, these different percentages of degradation can be attributed to differences between the cultivars, especially regarding the content of 5-CQA in green coffee. Studies have shown differences in genetic cultivar mainly in the concentration of 5-CQA [30, 31].

A wide range of organic acids values was found in the cultivars after roasting. In highest concentration was found for 5-CQA (0.82 to 1.93 g 100^{-1}) followed by quinic acid (0.66 to 1.05 g 100 g^{-1}), citric (0.52 to 0.79 g 100 g^{-1}) and malic (0.05 to 0.35 g 100 g^{-1}). Alcázar et al. [26] reported decrease of the content of citric acid (from 0.85 to 0.52 mg 100 g^{-1}) and malic acid (from 0.41 to 0.17 g 100 g^{-1}) in arabica coffees after roasting. Values of 1.15 and 1.67 g of 5-CQA 100 g^{-1} were observed in arabica commercial varieties of Brazilian coffee in weight loss of 13.6 and 14.1%, respectively. The decrease of citric acid (29 to 58% and an average of 54%) and malic acid (50 to 90% and an average of 43%) is shown in Figure 4. In the majority of the cultivars the malic acid were reduced to less than 60% (Bourbon, Catuaí, Icatu, Iapar 59, IPR 97, IPR 98, IPR 99, IPR 100, IPR 101, IPR 102, IPR 103 and IPR 104). The higher decrease of citric acid was observed for IPR 97 (29%) and the lower decrease for cultivar IPR 106 (58%).

Table 4. Pearson correlation between the degradation of 5-CQA and the formation of quinic acid in the cultivars

Variables*	QA G	5-CQA G	QA R	5-CQA R
QA G	1			
5-CQA G	-0.55	1		
Q A R	0.10	-0.41	1	
5-CQA R	-0.25	0.59	-0.46	1

Values in bold are different from 0 with a significance level alpha = 0.05.

* QA G – quinic acid of green coffee, QA R – quinic acid of roasted coffee, 5-CQA G – 5-CQA of green coffee and 5-CQA R – 5-CQA of roasted coffee.

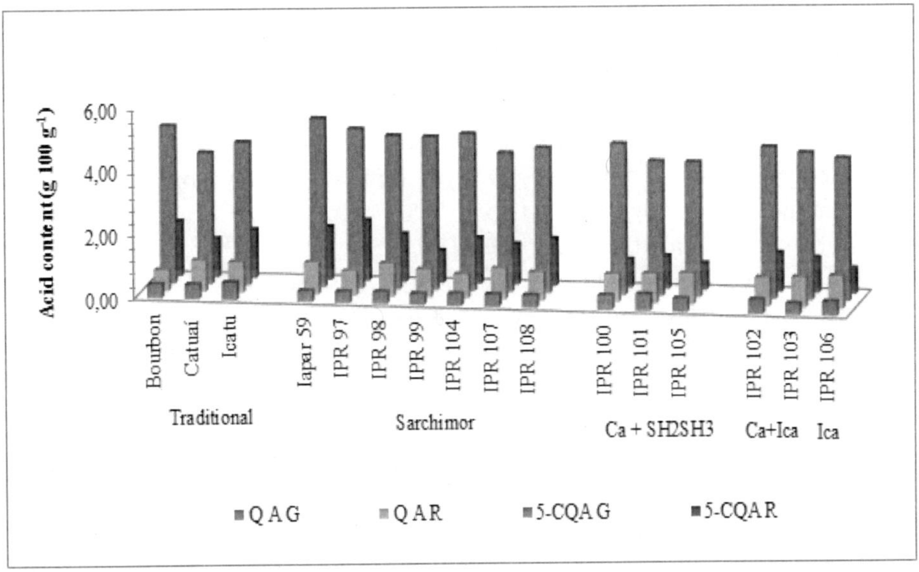

Figure 3. Degradation of 5-CQA and formation of quinic acid in the cultivars during roasting process.

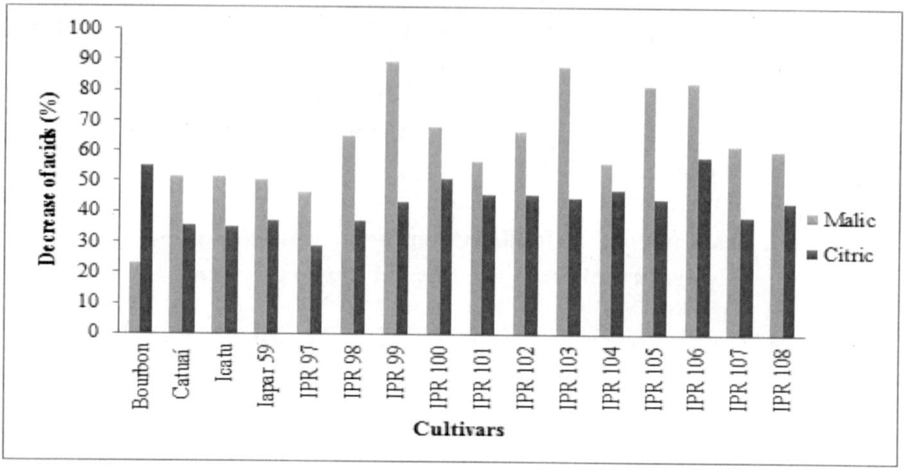

Figure 4. Percentages of decrease of citric acid and malic acids in coffee cultivars in roasting process.

Other acids such as acetic and lactic acid are also formed during roasting. The acetic and lactic acids appear to be generated during the roasting process using carbohydrates as precursors especially sucrose [6]. The cultivars evaluated in this study showed great variability in the formation of these acids ranging from 0.16 to 0.44 g 100^{-1}g (and an average 0.23 g 100^{-1}g) of latic acid and from 0.14 to 0.32 g 100^{-1}g (and an average 0.24 g 100^{-1}g) of acetic acid.

Generally, the acids are always formed during roasting in low concentrations, ranging from 0.005 to 0.17 g for the acetic acid and about 0.012 g of lactic acid 100 g^{-1} of coffee [26, 32].

The qualitative profile and the content of acids in the roasted coffee influence the acidity of the coffee brew [1]. Hydrolysis of esters and occurrence of the Maillard reaction during preparation are likely causes of increased acidity.

Coffees cultivars as Bourbon, Icatu and IPR 97, had the highest lactic acid formation (0.31, 0.44 e 0.34 g 100 g^{-1}). For acetic acid, the highest levels were observed for the cultivars Catuaí, IPR 98, IPR 102 e IPR 107 (0.32, 0.32, 0.29 e 0.32 g 100 g^{-1}, respectively) (Figure 5). The total titratable acidity could be influenced by several acids. The acids acetic, citric, malic have a great effect on the taste; high-molecular-weight acids and other acids have a minor contribution. Mineral acids as phosphoric acid also influence the coffee acidity [1, 5]. It can be over again observed that different contents of these acids can be attributed to differences between the cultivars, since all coffees were roasted in a similar way.

The pH values of coffee brews ranged from 4.98 to 5.35 and the acidity were between 2.55 to 3.33 mL. Brews of arabica and robusta coffee species subjected to medium roast showed pH of 4.98 and 5.24, respectively [33]. Several cultivars evaluated (Table 1) are derived from *C. canephora* via Sachimor, which possibly caused the decrease acidity. Cultivars of IPRs collection and traditional cultivars as Catuaí, Bourbon, Icatu and Tupi growing in others locals showed similar pH, that ranged of 5.12 to 5.24 and acidity value were ranged to 2.73 to 3.21 mL [34].

The action and participation of acids in the formation of perceived acidity in the coffees is not clearly understood. The consensus is that citric acid, malic acid, and acetic acid are important because they are present in high proportions and the pKa is similar to the coffee brews [6, 35, 36]. However, due to the buffering effects and the wide distributions of salts and acids present in coffee, it is difficult to predict the exact mechanism and which are the agents responsible for the perceived acidity in coffee, probably because many interactions are involved and act simultaneously in the formation of the brew acidity [35, 37].

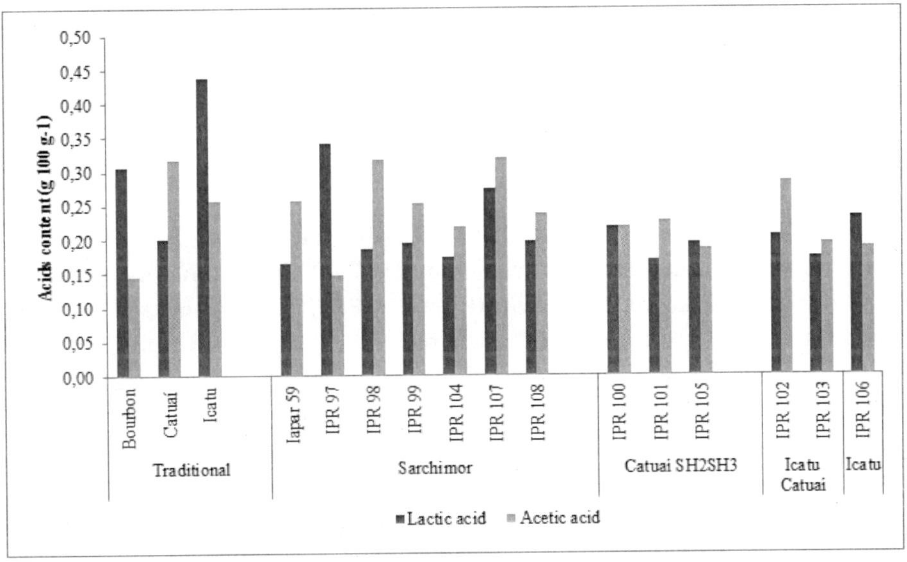

Figure 5. Content of lactic and acetic acids in roasted coffee cultivars.

Changes occurred during roasting for each cultivars can better observed by multivariate analysis such as principal component analysis (PCA) and hierarchical cluster analysis (HCA).

Principal Component Analysis (PCA) was used to describe the relationship between organic acids and coffee brew acidity of different cultivars. The first-two components accounted for 67% of the total variance. The first component (F1) was related to quinic acid (0.68), malic acid (-0.76), lactic acid (-0.59), 5-CQA (-0.84), pH (0, 89) and titratable acidity (-0.73). The second component (F2) was related to acetic acid (0.84) and citric acid (0.77). Bourbon, Icatu, IPR 97, IPR 104, IPR 108, IPR 102 and IPR 107 cultivars (left side of Figure 6a) were separated from the other cultivars by F1. These cultivars are characterized by high levels of lactic acid, malic acid, 5-CQA and titratable acidity. Catuaí, IPR 98, IPR 99, IPR 100, IPR 101, IPR 103, IPR 105 and IPR 106 cultivars have low values of these compounds (F1-) and high pH value and quinic acid. In this configuration, Iapar 59 is located in an intermediate position between the cultivars separated by F1.

Cultivars of the right and left superior quadrants of the figure are separated by acetic and citric acids content in F2 dimension. Catuaí, Icatu, Iapar 59, IPR 98, IPR 102, and IPR 107 were associated with higher citric and acetic acids contents (Figure 6a).

The Hierarchical Clustering Analysis (HCA) was also applied for grouping, considering the organic acids and the brew characteristics of roasted coffee. For roasted coffee, three groups in HCA were also formed (Figure 6b). The first group was composed by IPR 99, IPR 100, IPR 101, IPR 103, IPR 105 and IPR 106 with higher values of quinic, pH and intermediate values acetic and lower values of malic, citric, 5-CQA, lactic, and acidity (Table 5). The second group, formed by Bourbon and IPR 97, presented higher content of malic and latic acids, 5-CQA and acidity and intermediate content of citric and lower content of quinic, acetic acids and pH. The third group was composed by Catuaí, Icatu, Iapar 59, IPR 98, IPR 102, IPR 104, IPR 107 and IPR 108 that showed higher value of quinic, acetic, citric, and intermediate values of 5-CQA, lactic, malic acids, pH and acidity values (Table 5).

It can be verify that the total titratable acidity value did not differ between the groups but the contribution of the specific acids (acetic, malic and quinic acids mainly) for the formation of acidity was different. The acidity of G1 coffees presented high contribution of quinic acid, and lower contribution of malic and acetic acid. The acidity for G2 coffees was mainly associated to malic acid with a lower contribution of quinic and acetic acid. For G3 coffees the acidity was associated with quinic and acetic acid content. Lactic and citric acids has a similar contribution to acidity in all groups.

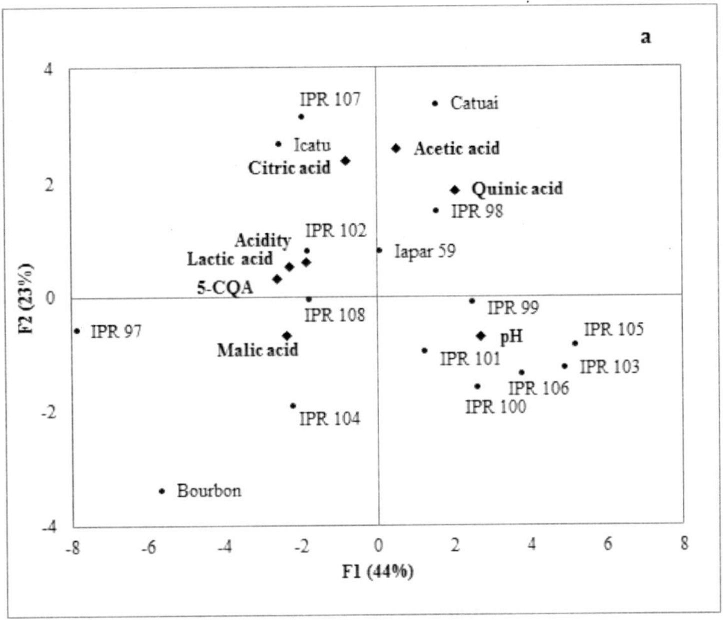

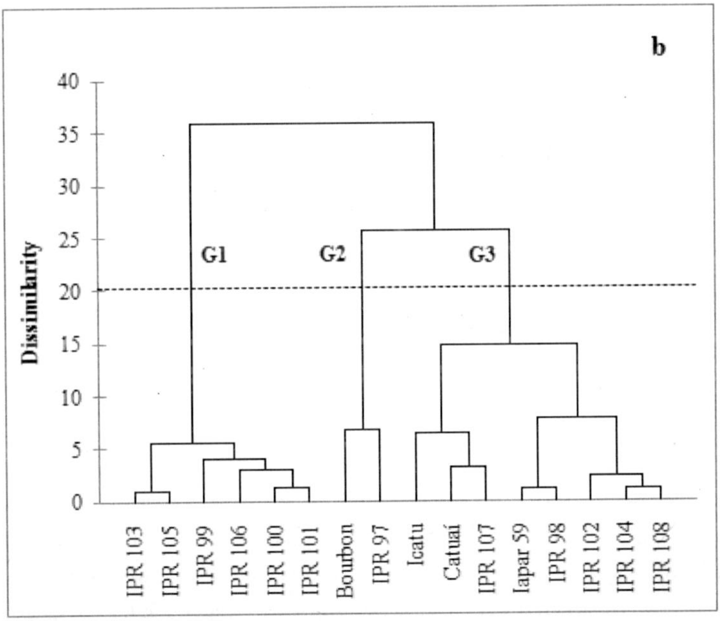

Figure 6. PCA Biplot (a) and HCA (b) of the coffee cultivars considering the acids profile and brews characteristics for roasted coffees.

Table 5. Contents of acids (g 100 g^{-1}), pH and acidity value for each HCA group of roasted coffees

Compounds	G1	G2	G3
Quinic	0.95 ± 0.04a	0.71 ± 0.06b	0.95 ± 0.08a
Malic	0.12 ± 0.01b	0.30 ± 0.05a	0.20 ± 0.03ab
Lactic	0.20 ± 0.02a	0.32 ± 0.02a	0.23 ± 0.09a
Acetic	0.21 ± 0.01b	0.14 ± 0.00b	0.28 ± 0.04a
Citric	0.58 ± 0.06a	0.63 ± 0.04a	0.65 ± 0.08a
5-CQA	0.97 ± 0.08c	1.84 ± 0.09a	1.42 ± 0.15b
pH	5.26 ± 0.05b	5.03 ± 0.05a	5.12 ± 0.04b
Acidity *	2.74 ± 0.14a	3.04 ± 0.29a	2.94 ± 0.17a

* mL of NaOH to 100 mL. Different small letters in the same line indicate significant differences among groups ($p > 0.05$).

Concerning the genetic background of cultivars, the G1 group was composed by the Catuaí SH2SH3 derived while the G3 group was composed mainly by Sarchimors and traditional cultivars. The Bourbon and IPR 97 cultivars participated in the distinct groups. It is also possible to observed that Icatu cultivars (IPR 106), Catuaí SH2SH3 derived (IPR 100, 101 and 105) and Sarchimor derived (Iapar 59, IPR 98, 104 and 108) remain grouped together even after the roasting (Figure 2 and Figure 6b) indicating a similarly influence of the roasting process. Catuaí cultivars derived presented a decrease in malic acid content, and Sarchimors an increased in the formation of quinic acid.

In this study, the cultivars showed different profiles of acids formed or degraded during roasting process. Further studies were required to clarify how the chlorogenic acids and other acids are influenced by roasting and in which way each acid contributed for the final acidity of the brew.

CONCLUSION

The green coffee beans of each cultivar showed diversity in the composition of organic acids and 5-CQA, but different profiles were observed after roasting. The formation of acid during the roasting was variable. During roasting the degradation of organic acids and 5-CQA and the formation of acetic, lactic and quinic acids followed different patterns in each cultivar.

Catuaí and Catuaí SH2SH3 showed a greater degradation of 5-CQA than Sarchimors derived cultivars.

In Catuaí and SH2SH3 derived the citric acid had the higher contribution for the formation of brew acidity. In the other hand, the acidity in Sarchimors and traditional cultivars was mainly related to the contribution of quinic and acetic acids.

These results suggested that the genetic background of each cultivar influenced the organic acids profile and the brews characteristics of the coffee brews.

REFERENCES

[1] Verardo, G.; Cecconi, F.; Geatti, P.; Giumanini, A.G., New procedures for determination of acids in coffee extracts, and observations on the development of acidity upon ageing, *Anal Bioanal Chem*. 2002, 374(5), 879-885.

[2] Rogers, W.J.; Michaux, S.; Bastin, M.; Bucheli, P., Changes to the content of sugars, sugar alcohols, myo-inositol, carboxylics acids and inorganic anions in developing grains from different varieties of Robusta (*Coffea canephora*) and Arabica (*C. arabica*) coffees, *Plant Sci.* 1999, 149(2) 115-123.

[3] Bhumiranata, N.; Adhikari, K.; Chambers, E., The development of an emotion lexicon for the coffee drinking experience, *Food Res Int.* 2014, 61, 83-92.

[4] Seo, H-S.; Lee, S-Y.; Hwang, I., Development of sensory attribute pool of brewed coffee, *J Sens Stud.* 2009, 24(1), 111-132.

[5] Balzer, H.H., Acids in coffee, In: R.J. Clarke, O.G. Vitzthum (Eds.), *Coffee Recent Developments*, Ed. Blackwell Science, Oxford, 2001, p. 18-28.

[6] Ginz, M.; Balzer, H.H.; Bradbury, A.G.W.; Maier, H.G., Formation of aliphatic acids by carbohydrate degradation during roasting of coffee, *Eur Food Res Technol*. 2000, 211(6), 404-410.

[7] Farah, A.; De Paulis, T.; Trugo, L.C.; Martin, P.R., Effect of roasting on the formation of chlorogenic acid lactones, *J Agric Food Chem*. 2005, 53 (5), 1505-1513.

[8] Farah, A.; Monteiro, M.C.; Calado, V.; Franca, A.S.; Trugo, L.C., Correlation between cup quality and chemical attributes of Brazilian coffee, *Food Chem*. 2006, 98(2), 373-380.

[9] Ping, W.; Rong, Z., Determination of Organic Acids Exuded from Plant Roots by High Performance Liquid Chromatography, *Chin J Chromatogr*. 2006, 24(3), 239-242.

[10] Ito, D.S.; Sera, T.; Sera, G.H.; Del Grossi, L.; Kanayama, F.S., Resistance to bacterial blight in arabica coffee cultivars, *Crop Breed Appl Biot*. 2008, 8, 99-103.

[11] Sera, G.H.; Sera, T.; Ito, D.S.; Mata, J.S.; Doi, D.S.; Azevedo, J.A.; Ribeiro Filho, C., Progênies de *Coffea arabica* cv IPR-100 resistentes ao nematóide *Meloidogyne paranaensis* (Coffea arabica progenies IPR- 100 hp resistant to Meloidogyne paranaensis). *Bragantia*. 2007, 6(1), 43-49.

[12] Santos, J.C.F., Origens e características dos cultivares de café, 2011 (Origins and characteristics of coffee cultivars, 2011). http://www.revistacafeicultura.com.br/index.php?tipo=ler&mat=3696html (accessed November, 2015).

[13] Fazuoli, L.C., Melhoramento de *Coffea arabica* - Melhoramento do café - Mesa redonda, Centro de Café 'Alcides Carvalho' – IAC (Coffea arabica improvement - Coffee Improvement - Round Table, Coffee Centre 'Alcides Carvalho' – IAC), *Guaraparí*, 11-08-2009, 2009.

[14] Eira, M.T.S.; Fazuoli, L.C.; Guerreiro Filho, O.; Silvarolla, M.B.; Ferrão, M.A.G.; Fonseca, A.F.A.; Ferrão, R.G.; Sera, T.; Pereira, A.A.; Sakiyama, N.S.; Zambolim, L.; Carvalho, C.H.; Padilha, L.; Souza, F.F., Bancos de germoplasma de café no Brasil: base do melhoramento para produtividade e qualidade (Banks coffee germplasm in Brazil: base improvement for productivity and quality). In: V Simpósio de Pesquisa dos Cafés do Brasil, Águas de Lindóia, SP, Brazil, 5-11 maio 2007, p. 5. Brasília: Embrapa – Café, 2007.

[15] Sera, T.; Mata, J.S.; Sera, G.H.; Doi, D.S.; Ito, D.S.; Azevedo, J.A.; Ribeiro Filho, C., Identificação de novas progênies da cultivar IPR 100 resistentes ao nematóide *Meloidogyne paranaensis* (Identification of new progenies of cultivar IPR 100 resistant to Meloidogyne paranaensis). In: *IV Simpósio de Pesquisa dos Cafés do Brasil* (IV Research Symposium of Coffees of Brazil), Londrina, PR, Brazil, 2-5 maio, 2005, p. 5. Brasília: Embrapa Café, 2005.

[16] Alteia, M.Z.; Sera, T.; Guerreiro, A.; Azevedo, J.A.; Colombo, L.A., Vinte anos de avaliações agronômicas das progênies dos germoplasmas: Catimor, Sarchimor, "Icatu"x'Catuaí'e Catuaí x Sh1, Sh2, Sh3, Sh4 em Londrina-Paraná (Twenty years of agronomic evaluations of the progenies of germplasm: Catimor, Sarchimor, "Icatu" x'Catuaí'e Catuaí x Sh1, Sh2, Sh3, Sh4 in Londrina, Paraná). In: *II Simpósio de Pesquisa*

dos Cafés do Brasil (II Research Symposium of Coffees of Brazil), Vitória, Brazil, 24-27 setembro 2001, pp. 1412-1420. Brasília: Embrapa Café. 2001.

[17] Kitzberger, C.S.G.; Scholz, M.B.S.; L.F.P.; Pereira, Benassi, M.T., Composição química de cafés árabica de cultivares tradicionais e modernas (Chemical composition of Arabica coffees from traditional and modern cultivars). *Pesq Agropec Bras*. 2013, 48(11), 1498-1506.

[18] Alves, S.T.; Dias, R.C.E.; Benassi, M.T.; Scholz, M.B.S., Metodologia para análise simultânea de ácido nicotínico, trigonelina, ácidos clorogênicos e cafeína em café torrado por cromatografia líquida de alta eficiência (Methodology for simultaneous analysis of nicotinic acid, trigoneline, chlorogenic acid and caffeine in roasted coffee by high-performance liquid chromatography). *Quim. Nova*. 2006, 29(6), 1146-1148.

[19] AddinsoftXLStat: software for statistical analysis. *Versão* 2007.8. Paris, 2007. 1 CD-ROM.

[20] Steiman, S.R., Method development for green coffee analysis and its possible application for group discrimination and correlation of green coffee chemistry with cupping quality. MS thesis, University Hawaii Library, 2003.

[21] Koshiro, Y.; Jackson, M.C.; Katahira, R.; Wang, M-L.; Nagai, C.; Ashihara, H., Biosynthesis of Chlorogenic Acids in Growing and Ripening Fruits of *Coffea arabica* and *Coffea canephora* Plants, *Z. Naturforsch*. 2007, 62(9-10), 731-742.

[22] Vaast, P.; Bertrand, B.; Perriot, J-J.; Guyot, B.; Génard, M., Fruit thinning and shade improve bean characteristics and beverage quality of coffee (*Coffea arabica* L.) under optimal conditions, *J. Sci. Food Agr.* 2006, 86(2), 197-204.

[23] Farah, A.; Donangelo, C.M., Phenolic compounds in coffee, *Braz J Plant Physio*. 2006, 18(1), 23-36.

[24] Jöet, T.; Laffargue, A.; Salmona, J.; Doulbeau, S.; Descroix, F.; Bertrand, B.; Kochko, A.; Dussert, S., Metabolic pathways in tropical dicotyledonous albuminous seeds: Coffea arabica as a case study. *New Phytol.* 2009, 182(1), 146-162.

[25] Rodrigues, C.I.; Marta, L.; Maia, R.; Miranda, M.; Ribeirinho, M.; Máguas, C., Application of solid-phase extraction to brewed coffee caffeine and organic acid determination by UV/HPLC, *J Food Compos Anal.* 2007, 20(5), 440-448.

[26] Alcázar, A.; Fernández-Cáceres, P.L.; Martín, M.J.; Pablos, F.; González, A.G., Ion chromatographic determination of some organic acids, chloride and phosphate in coffee and tea, *Talanta.* 2003, 61(2) 95-101.

[27] Moon, J.; Yoo, H.I.; Shibamoto, T., Role of Roasting Conditions in the Level of Chlorogenic Acid Content in Coffee Beans: Correlation with Coffee Acidity, *J Agr Food Chem.* 2009, 57(12) 5365-5369.

[28] Scholz, M.B.; Maier, G.H., Isomers of quinic acid and quinide in roasted coffee, *Z Lebensm Unters Forsch.* 1990, 190, 132-134.

[29] Leloup, V.; Louvrier, A.; Liardon, R., Degradation mechanisms of chlorogenic acids during roasting. In: *16th International Scientific Colloquium Coffee*, Kyoto, Japan, 9-14 abril, 1995, p. 192-198, Association Scientifique Internationale du Café (International Scientific Coffee Association), 1995.

[30] Guerrero, G.; Suarez, M.; Moreno, G., Chlorogenic acids as potential criterion in coffee genotype selection, *J Agric Food Chem.* 2001, 49(5), 2454-2458.

[31] Bicho, N.C.; Leitão, A.E.; Ramalho, J.C.; Alvarenga, N.B.; Lidon, F.C., Chemical descriptors for sensory and parental origin of commercial Coffea genotypes. *Int J Food Sci Nutr.* 2012, 62(7), 865-871.

[32] Galli, V.; Barbas, C., Capillary electrophoresis for the analysis of short-chain organic acids in coffee, *J. Chromatogr. A.* 2004, 1032(1-2), 299-304.

[33] Bicho, N.C.; Leitão, A.E.; Ramalho, J.C.; Alvarenga, N.B.; Lidon, F.C., Identification of nutritional descriptors of roasting intensity in beverages of Arabica and Robusta coffee beans, *Int J Food Sci Nutr.* 2011, 62(8), 865-871.

[34] Scholz, M.B.S.; Silva, J.V.N.; Figueiredo, V.R.G.; Kitzberger, C.S.G., Atributos sensoriais e características físico-químicas de bebida de cultivares de café do Iapar (sensory attributes and physico-chemical characteristics of drinking Iapar coffee cultivars), *Coffee Sci.* 2013, 8(1), 6-16.

[35] Neta, E.R.; Johanningsmeier, S.D.; McFeeters, R.F., The Chemistry and physiology of sour taste—A Review, *J Food Sci.* 2007, 72(2), R33-R38.

[36] Neta, E.R.; Johanningsmeier, S.D.; Drake, M.A.; McFeeters, R.F., Effects of pH adjustment and sodium ions on sour taste intensity of organic acids, *J Food Sci.* 2009, 74(4), 165-169.

[37] Glöss, A.; Edelmann, B.; Klopprogge, B.; D'Ambrosio, L.; Chatelain, K.; Bongartz, A.; Strittmatter, A.; Rast, M.; Yeretzian, C., Comparison

of nine common coffee extraction methods: instrumental and sensory analysis, *Eur Food Res Technol.* 2013, 236(4), 607-627.

Biographical Sketch

Cíntia Sorane Good Kitzberger

Affiliation: Instituto Agronômico do Paraná (IAPAR)

Education: BSc in Food Engineering (Universidade Norte do Paraná, 1998), MSc in Food Engineering (Universidade Federal de Santa Catarina, 2005) and Ph.D. in Food Science (Universidade Estadual de Londrina, 2012).

Address: Área de Ecofisiologia – Laboratório de Fisiologia Vegetal, IAPAR - Instituto Agronômico do Paraná, Rodovia Celso Garcia Cid, km 375 86047-902 - Londrina – PR, Brazil
Phone number: +55 (43) 33762000; Fax number: +55 (43) 33762101
URL: http://www.iapar.br/
Contact e-mail: cintia_kitzberger@iapar.br
alternative e-mail: cintiasorane@yahoo.com.br

Research and Professional Experience:

Cíntia Sorane Good Kitzberger, Research Assistant at IAPAR, has experience in Food Science and Technology, focusing on Chemistry and Biochemistry of Foods. Her field of expertise covers the following subjects: coffee, wheat, fruits, beans, technological properties of different cultivars, antioxidant activity, HPLC and sensory analysis.

Regarding the research grants on the last 10 years, she was participated in three projects as a team member: 1 project supported by the Institute Agronomic of Paraná/IAPAR (2010-2013), 1 project supported by the Brazilian National Council for Scientific and Technological Development (CNPq) (2010-2012) and 1 project supported by the Consórcio Pesquisa Café (2014-current).

She is the author of 13 articles published in scientific journal; 48 articles published in events proceedings. She acts as a reviewer for some scientific journal (Journal of Agricultural and Food Chemistry, Crop Science and Technology, Semina. Ciências Agrárias, Journal of Food Composition and

Analysis, African Journal of Food Science and African Journal of Agricultural Research).

Regarding the scientific impact (until march 2016), her publications received 232 citations with h-index 5 at Google Scholar, and 125 citations and h-index 4 at Web of Science.

Address to access CV: http://lattes.cnpq.br/8650340803878846

Research ID D-9635-2046

Professional Appointments:

Instituto Agronômico do Paraná – Department of Ecophysiology (2006 - current) – Researcher assistant.

Serviço Nacional de Aprendizagem – (2005-2005) - professor

Publications Last 3 Years:

Articles:

Conti, M.C.M.D., Kitzberger, C.S.G., Scholz, M.B.S., Prudencio, S.H. Características físicas e quimicas de cafés torrados e moídos exóticos e convencionais (Physical and chemical characteristics of roasted coffee and exotic ground and conventional). *Boletim do Centro de Pesquisa e Processamento de Alimentos* (Bulletin Research Center and Food Processing), v. 31, p. 161-172, 2013.

Kitzberger, C.S.G., Benassi, M.T., Scholz, M.B.S., Pereira, L.F.P. Composição química de cafés árabica de cultivares tradicionais e modernas (Chemical composition of Arabica coffees from traditional and modern cultivars). *Pesquisa Agropecuária Brasileira* (Brazilian Agricultural Research), v. 48, p. 1498-1506, 2013.

Kitzberger, C.S.G., Scholz, M.B.S., Pereira, L.F.P., Vieira, L.G.E., Sera, T., Silva, J.B.G.D., Benassi, M.T. Diterpenes in green and roasted coffee of Coffea arabica cultivars growing in the same edapho-climatic conditions. *Journal of Food Composition and Analysis*, v. 30, p. 52-57, 2013.

Kitzberger, C.S.G., Scholz, M.B.S., Benassi, M.T. Bioactive compounds content in roasted coffee from traditional and modern Coffea arabica cultivars grown under the same edapho-climatic conditions. *Food Research International*, v. 61, p. 61-66, 2014.

Scholz, M.B.S.; Kitzberger, C.S.G.; Pagiatto, N.F.; Pereira, L.F.P.; Davrieux, F.; Pot, D.; Charmetant, P.; Leroy, T. Chemical composition in wild ethiopian Arabica coffee accessions. *Euphytica* (Wageningen), p. 1-10, 2016. DOI 10.1007/s10681-016-1653-y.

Scholz, M.B.S.; Kitzberger, C.S.G.; Pagiatto, N.F.; Pereira, L.F.P.; Davrieux, F.; Pot, D.; Charmetant, P.; Leroy, T. Application of near infrared spectroscopy for green coffee biochemical phenotyping. *Journal of Near Infrared Spectroscopy*, v. 22, p. 411-421, 2014.

Scholz, M.B.S., Pagiatto, N.F., Kitzberger, C.S.G., Pereira, L.F.P., Davrieux, F., Charmetant, P., Leroy, T. Validation of near-infrared spectroscopy for the quantification of cafestol and kahweol in green coffee. *Food Research International*, v. 61, p. 176-182, 2014.

Scholz, M.B.S., Silva, J.V.N., Figueiredo, V.R.G., Kitzberger, C.S.G. Atributos sensoriais e características físico-químicas de bebida de cultivares de café do Iapar (sensory attributes and physico-chemical characteristics of drinking Iapar coffee cultivars). *Coffee Science*, v. 8, p. 6-16, 2013.

Maria Brígida dos Santos Scholz

Affiliation: Instituto Agronômico do Paraná (IAPAR)

Education: BSc in Pharmacy and Biochemistry (Universidade Federal do Paraná, 1977), MSc in Food Science (Universidade Estadual de Londrina, 1990), Ph.D. in Food Science (Universidade Estadual de Londrina, 2007) and Postdoctoral (La Recherche Agronomique pour le Développement, CIRAD, França, 2013).

Address:

Área de Ecofisiologia – Laboratório de Fisiologia Vegetal, IAPAR - Instituto Agronômico do Paraná, Rodovia Celso Garcia Cid, km 375 86047-902 - Londrina – PR, Brazil

Phone number: +55 (43) 33762000; Fax number: +55 (43) 33762101

URL: http://www.iapar.br/

Contact e-mail: mbscholz@iapar.br

alternative e-mail: mbrischolz@hotmail.com

Research and Professional Experience:

Maria Brígida dos Santos Scholz, researcher at IAPAR, has experience in Food Science and Technology, focusing on Chemistry and Biochemistry of Foods. Her field of expertise covers the following subjects: coffee, wheat, fruits, beans, honey, and technological properties of different cultivars, sensory analysis and NIRs.

She advised 2 Master's thesis and 4 Ph.D. thesis, and 2 Post-doctoral supervisions. She currently supervises 2 Master's thesis, 2 Ph.D. thesis and 2 Post-doctoral researchers.

Regarding the research grants on the last 10 years, she was responsible for six projects as research leader (3 projects supported by the Brazilian National Council for Scientific and Technological Development (CNPq), 1 project supported by Secretaria de Estado da Ciência, Tecnologia e Ensino Superior and 2 projects supported by the Consórcio Pesquisa Café.

She is the author of 44 articles published in scientific journal; 3 book chapters; 107 articles published in events proceedings. She acts as a reviewer for some scientific journal (Bragantia, Ciência e Agrotecnologia, Scientia Agraria, Revista Ciência e Tecnologia de Alimentos, Food Research International).

Regarding the scientific impact (until march 2016), her publications received 469 citations with h-index 11 at Google Scholar, and 371 citations and h-index 10 at Web of Science.

Address to access CV: http://lattes.cnpq.br/2593214529990790

Research ID E-7958-2014

Professional Appointments:

Instituto Agronômico do Paraná (1991 - current): researcher

Universidade Estadual de Londrina (1989-1991): professor.

Publications Last 3 Years:

Articles:

Costa, M.S.; Scholz, M.B.S.; Franco, C.M.L. Effect of high and low molecular weight glutenin subunits, and subunits of gliadin on physicochemical parameters of different wheat genotypes. *Ciência e Tecnologia de Alimentos (Impresso)* (Food Science and Technology (Printed)), v. 33, p. 163-170, 2013.

Conti, M.C.M.D., Kitzberger, C.S.G., Scholz, M.B.S., Prudencio, S.H. Características físicas e quimicas de cafés torrados e moídos exóticos e convencionais (Physical and chemical characteristics of roasted coffee and exotic ground and conventional). *Boletim do Centro de Pesquisa e Processamento de Alimentos* (Bulletin Research Center and Food Processing), v. 31, p. 161-172, 2013.

Kitzberger, C.S.G., Benassi, M.T., Scholz, M.B.S., Pereira, L.F.P. Composição química de cafés árabica de cultivares tradicionais e modernas (Chemical composition of Arabica coffees from traditional and modern cultivars). *Pesquisa Agropecuária Brasileira* (Brazilian Agricultural Research), v. 48, p. 1498-1506, 2013.

Kitzberger, C.S.G., Scholz, M.B.S., Pereira, L.F.P., Vieira, L.G.E., Sera, T., Silva, J.B.G.D., Benassi, M.T. Diterpenes in green and roasted coffee of Coffea arabica cultivars growing in the same edapho-climatic conditions. *Journal of Food Composition and Analysis*, v. 30, p. 52-57, 2013.

Kitzberger, C. S. G., Scholz, M. B. S., Pereira, L. F. P., Vieira, L. G. E., Sera, T., Silva, J. B. G. D., Benassi, M. T. Diterpenes in green and roasted coffee of Coffea arabica cultivars growing in the same edapho-climatic conditions. *Journal of Food Composition and Analysis*, v. 30, p. 52-57, 2013 (http://dx.doi.org/10.1016/j.jfca.2013.01.007).

Kitzberger, C.S.G.; Scholz, M.B.S.; Benassi, M.T.Bioactive compounds content in roasted coffee from traditional and modern Coffea arabica cultivars grown under the same edapho-climatic conditions. *Food Research International*, v. 61, p. 61-66, 2014.

Link, J.V.; Lemes, A.L.G.; Marquetti, I.; Scholz, M.B.S.; Bona, E. Geographical and genotypic classification of arabica coffee using fourier transform infrared spectroscopy and radial-basis function networks. *Chemometrics and Intelligent Laboratory Systems*, v. 135, p. 150-154, 2014.

Link, J.V.; Lemes, A.L.G.; Marquetti, I.; Scholz, M.B.S.; Bona, E. Geographical And Genotypic Segmentation Of Arabica Coffee Using Self-Organizing Maps. *Food Research International*, v. 59, p. 1-7, 2014.

Marquetti, I.; Link, J.V.; Lemes, A.L.G.; Scholz, M.B.S.; Valderrama, P.; Bona, E. Partial least square with discriminant analysis and near infrared spectroscopy for evaluation of geographic and genotypic origin of arabica coffee. *Computers and Electronics in Agriculture*, v. 121, p. 313-319, 2016.

Riede, C.R.; Campos, L.A.C.; Okuyama, L.A.; Scholz, M.B.S.; Shioga, P.S.; Pola, J.N.; Machado, J.C.; Garbuglio, D.D. IPR Catuara TM - New cultivar of high gluten wheat. *Crop Breeding and Applied Biotechnology*, v. 15, p. 56-58, 2015.

Scholz, M.B.S.; Pagiatto, N.F.; Kitzberger, C.S.G.; Pereira, L.F.P.; Davrieux, F.; Charmetant, P.; Leroy, T. Validation of near-infrared spectroscopy for the quantification of cafestol and kahweol in green coffee. *Food Research International*, v. 61, p. 176-182, 2014.

Scholz, M.B.S.; Kitzberger, C.S.G.; Pereira, L.F.P.; Davrieux, F.; Pot, D.; Charmetant, P.; Leroy, T. Application of near infrared spectroscopy for green coffee biochemical phenotyping. *Journal of Near Infrared Spectroscopy*, v. 22, p. 411, 2014.

Scholz, M.B.S.; Silva, J.V.N.; Figueiredo, V.R.G.; Kitzberger, C.S.G. Atributos sensoriais e características físico-químicas de bebida de cultivares de café do Iapar (sensory attributes and physico-chemical characteristics of drinking Iapar coffee cultivars). *Coffee Science*, v. 8, p. 6-16, 2013.

Book Chapter:

Scholz, M.B.S., Takahashi, M. Utilização da mandioca (Cassava). In: Mário Takahashi, Nelson da Silva Fonseca Júnior, Sônia Martins Torrecillas (Org.). *Mandioca - antes, agora e sempre* (Cassava - before, now and forever). Londrina-PR: 2002, p. 19-41.

Scholz, M.B.S., Qualidade Tecnológica do Arroz. Arroz Irrigado - Práticas de Cultivo (Technological rice quality. Irrigated Rice - Farming Practices). Londrina -PR: Iapar, 2001, p. 190-197.

Scholz, M.B.S., Qualidade Tecnológica de Variedades de Feijão (Technological quality bean varieties). Feijão - Tecnologia de Produção (Beans - Production Technology). Londrina -PR: Iapar, 2000, p. 101-108.

Marta de Toledo Benassi

Affiliation: Universidade Estadual de Londrina (UEL)

Education: BSc in Food Engineering (Universidade Estadual de Campinas,1985), MSc in Food Science (Universidade Estadual de Campinas, 1990) and Ph.D. in Food Science (Universidade Estadual de Campinas, 1997).

Address:

Departamento de Ciência e Tecnologia de Alimentos/Centro de Ciências Agrárias/Universidade Estadual de Londrina

86057970 - Londrina, PR - Brasil

Phone number: +55 (43) 33715970; Fax number: +55 (43) 33714080

URL: http://www.uel.br/cca/dcta

Contact e-mail: martatb@uel.br;

alternative e-mail: marta.benassi@gmail.com

Research and Professional Experience:

Marta de Toledo Benassi, professor at UEL, has experience in Food Science and Technology, focusing on Chemistry and Biochemistry of Foods. Her field of expertise covers the following subjects: coffee, antioxidant activity, HPLC, sensory analysis.

She advised 15 Master's thesis and 4 Ph.D. thesis, and 2 Post-doctoral supervisions. She currently supervises 2 Master's thesis, 2 Ph.D. thesis and 2 Post-doctoral researchers.

Regarding the research grants on the last 10 years, she was responsible for eight projects as research leader (3 projects supported by the Brazilian National Council for Scientific and Technological Development (CNPq), 4 projects supported by Universidade Estadual de Londrina, 1 project supported by the Brazilian Agricultural Research Corporation (EMBRAPA)) and participated in six projects as a team member (4 projects supported by the Brazilian National Council for Scientific and Technological Development (CNPq), 1 project supported by the Coordination for the Improvement of Higher Education Personnel (CAPES), 1 project supported by the São Paulo Research Foundation (FAPESP)). Regarding ongoing external funding, she is responsible for one project as research leader (Assessment of composition, sensory characteristics and cup quality of *Coffea canephora* brews (2014-2017), supported by CNPq) and participate in one project as team member (Brazilian pine (*Araucaria angustifolia*): assessment Brazilian pine: assessment of potential in the feed and in the development of new products (2012-2016), supported by EMBRAPA).

She was awarded with Scientific Productivity grants funded by the Paraná State Funding Agency/Fundação Araucária (2009-2010) and CNPq (2011-2013, 2014-current)

She is the author of 79 articles published in scientific journal; 4 book chapters; 167 articles published in events proceedings. She was also member of Editorial Board of the journal Semina. Ciências Agrárias (2011-2014) and acts as a reviewer for several scientific journal (Journal of Sensory Studies; Food Research International; Journal of Food Composition and Analysis; Journal of Agricultural and Food Chemistry; Food Chemistry; Lebensmittel-Wissenschaft + Technologie).

Regarding the scientific impact (until march 2016), her publications received 1145 citations with h-index 18 at Google Scholar, and 415 citations and h-index 10 at Web of Science.

Address to access CV: http://lattes.cnpq.br/7409756675845441

Research ID F-7213-2012

Professional Appointments:

Universidade Estadual Paulista Júlio de Mesquita Filho – UNESP (1991-1997): professor

Universidade Estadual de Londrina (1998 - current): Associate-professor.

Management and Administrative Positions: Deputy Head of Food Science and Technology Department (2006-2008, 2008-2010), Coordinator of Food Science Graduate Program (MS and PhD) (2013-current)

Publications Last 3 Years:

Articles:

Chisté, R. C., Mercadante, A. Z., Benassi, M. T. Efficiency of different solvents on the extraction of bioactive compounds from the amazonian fruit *Caryocar villosum* and the effect on its antioxidant and colour properties. *Phytochemical Analysis*, v. 25, p. 364-372, 2014 (http://dx.doi.org/10.1002/pca.2489).

Corso, M. P., Benassi, M.T. Packaging attributes of antioxidant-rich instant coffee and their influence on the purchase intent. *Beverages*, v. 1, p. 273-291, 2015 (http://dx.doi.org/10.3390/beverages1040273).

Corso, M. P.; Vignoli, J. A.; Benassi, M.T. Development of an instant coffee enriched with chlorogenic acids. *Journal of Food Science and Technology*, v. 53, p. 1-9, 2016 (http://link.springer.com/10.1007/s13197-015-2163-y).

Dias, R. C. E., Benassi, M.T. Discrimination between arabica and robusta coffees using hydrosoluble compounds: Is the efficiency of the parameters dependent on the roast degree? *Beverages*, v. 1, p. 127-139, 2015 (http://dx.doi.org/10.3390/beverages 1030127).

Dias, R. C. E., Faria, A. F., Bragagnolo, N., Mercadante, A. Z., Benassi, M.T. Roasting process affects the profile of diterpenes in coffee. *European Food Research and Technology*, v. 239, p. 961-967, 2014 (http://dx.doi.org/10.1007/s00217-014-2293-x).

Dias, R. C. E., Faria, A. F., Mercadante, A. Z., Bragagnolo, N., Benassi, M. T. Comparison of extraction methods for kahweol and cafestol analysis in roasted coffee. *Journal of the Brazilian Chemical Society*, v. 24, p. 492-499, 2013 (http://dx.doi.org/10.5935/0103-5053.20130057).

Dias, R. C. E., Alves, S. T., Benassi, M.T. Spectrophotometric method for quantification of kahweol in coffee. *Journal of Food Composition and Analysis*, v. 31, p. 137-143, 2013 (http://dx.doi.org/10.1016/j.jfca.2013.04.001).

Francisco, J. S., Santos, A. C. F., Benassi, M.T. Efeito das informações e características da embalagem na expectativa e aceitação de café solúvel adicionado de café torrado micronizado (Effect of information and package characteristics in expectation and soluble coffee acceptance added micronized roasted coffee). *Brazilian Journal of Food Technology*, v. 17, p. 243-251, 2014 (http://dx.doi.org/10.1590/1981-6723.1614).

Godoy, R.; Caliari, M.; Soares Junior, M. S.; Silva, V. S. N.; Benassi, M.T.; Garcia, M. C. Quinoa and Rice Co-products Gluten Free-cereals: Physical, Chemical, Microbiological and Sensory Qualities. *Journal of Food and Nutrition Research*, v. 3, p. 599-606, 2015 (www.sciepub.com/.../downloads?doi=10.12691/jfnr-3-9-7).

Kitzberger, C. S. G., Scholz, M. B. S., Benassi, M. T. Bioactive compounds content in roasted coffee from traditional and modern *Coffea arabica* cultivars grown under the same edapho-climatic conditions. *Food Research International*, v. 61, p. 61-66, 2014 (http://dx.doi.org/10.1016/j.foodres.2014.04.031).

Kitzberger, C. S. G., Scholz, M. B. S., Pereira, L. F. P., Benassi, M.T. Composição química de cafés árabica de cultivares tradicionais e modernas (Chemical composition of Arabica coffees from traditional and modern cultivars). *Pesquisa Agropecuária Brasileira* (Brazilian Agricultural Research), v. 48, p. 1498-1506, 2013 (http://dx.doi.org/10.1590/S0100-204X2013001100011).

Kitzberger, C. S. G., Scholz, M. B. S., Pereira, L. F. P., Vieira, L. G. E., Sera, T., Silva, J. B. G. D., Benassi, M. T. Diterpenes in green and roasted coffee of *Coffea arabica* cultivars growing in the same edapho-climatic conditions. *Journal of Food Composition and Analysis*, v. 30, p. 52-57, 2013 (http://dx.doi.org/10.1016/j.jfca.2013.01.007).

Kobayashi, M. L., Benassi, M. T. Impact of packaging characteristics on consumer purchase intention: Instant coffee in refill packs and glass jars. *Journal of Sensory Studies*, v. 30, p. 1-12, 2015 (http://dx.doi.org/10.1111/joss.12142).

Mamede, M. E. O., Kalschne, D. L., Santos, A. P. C., Benassi, M.T. Cajá-flavored drinks: a proposal for mixed flavor beverages and a study of the consumer profile. *Food Science and Technology*, v. 35, p. 143-149, 2015. (http://dx.doi.org/10.1590/1678-457X.6563).

Marcucci, C. T., Almeida, M. B., Nixdorf, S. L., Benassi, M. T. Teores de trigonelina, ácido 5-cafeoilquínico, cafeína e melanoidinas em cafés solúveis comerciais brasileiros (Levels of trigoneline, 5-caffeoylquinic acid, caffeine and melanoidins in Brazilian commercial soluble coffees).

Química Nova, v. 36, p. 544-548, 2013 (http://dx.doi.org/10.1590/S0100-40422013000400011).

Poerner-Rodrigues, N., Benassi, M.T., Bragagnolo, N. Scavenging capacity of coffee brews against oxygen and nitrogen reactive species and the correlation with bioactive compounds by multivariate analysis. *Food Research International*, v. 61, p. 228-235, 2014 (http://dx.doi.org/10.1016/j.foodres.2013.09.028).

Rezende, N.V., Benassi, M.T., Grossmann, M. V. E. Effects of fat replacement and fibre addition on the texture, sensory acceptance and structure of sucrose-free chocolate. *International Journal of Food Science and Technology*, v. 50, p. 1413-1420, 2015 (http://dx.doi.org/10.1111/ ijfs.127 91).

Rezende, N.V., Benassi, Marta T., Vissotto, F.Z., Augusto, P.C., Grossmann, M. V. E. Mixture design applied for the partial replacement of fat with fibre in sucrose-free chocolates. Lebensmittel-Wissenschaft + Technologie/Food Science + Technology, v. 62, p. 598-604, 2015 (http:// dx.doi.org/10.1016/j.lwt.2014.08.047).

Sousa, J. M., Souza, E., Marques, G., Benassi, M. T., Gullón, B., Pintado, M. M., Magnani, M. Sugar profile, physicochemical and sensory aspects of monofloral honeys produced by different stingless bee species in Brazilian semi-arid region. Lebensmittel-Wissenschaft + Technologie/Food Science + Technology, v. 65, p. 645-651, 2016. (http://dx.doi.org/10.1016/ j.lwt. 2015.08.058).

Terhaag, M. M., Almeida, M. B., Benassi, M.T. Soymilk plain beverages: correlation between acceptability and physical and chemical characteristics. *Food Science and Technology*, v. 33, p. 387-394, 2013 (http://dx.doi.org/10.1590/S0101-20612013005000052).

Vignoli, J. A., Viegas, M. C., Bassoli, D. G., Benassi, M. T. Roasting process affects differently the bioactive compounds and the antioxidant activity of arabica and robusta coffees. *Food Research International*, v. 61, p. 279-285, 2014 (http://dx.doi.org/10.1016/j.foodres.2013.06.006).

Vissotto, L. C., Rodrigues, E., Chisté, R. C., Benassi, M.T., Mercadante, A. Z. Correlation, by multivariate statistical analysis, between the scavenging capacity against reactive oxygen species and the bioactive compounds from frozen fruit pulps. *Food Science and Technology*, v. 33, p. 57-65, 2013 (http://dx.doi.org/10.1590/s0101-20612013000 500010).

Book Chapter:

Benassi, M. T., DIAS, R. C. E. Assay of kahweol and cafestol in coffee In: *Coffee in Health and Disease Prevention*. 1 ed. London: Elsevier, 2015, v. 1, p. 993-1004 (http://dx.doi.org/10.1016/B978-0-12-409517-5.00109-1).

Benassi, M.T.; Corso, M. P. Effects of Extrinsic Factors on the Acceptance of Instant Coffee Enriched with Natural Antioxidants from Green Coffee (in press). In: John L. Massey (Org.). *Coffee: Production, Consumption and Health Benefits* (Series: Food and Beverage Consumption and Health). 1 ed. Hauppauge: Nova Publishers, 2016, in press.

Vignoli, J. A.; Viegas, M. C.; Bassoli, D. G.; Benassi, M.T. Coffee Brews Preparation: Extraction of Bioactive Compounds and Antioxidant Activity (in press). In: John L. Massey (Org.). *Coffee: Production, Consumption and Health Benefits* (Series: Food and Beverage Consumption and Health). 1 ed. Hauppauge: Nova Publishers, 2016, in press.

João Batista Gonçalves Dias da Silva

Affiliation: COCARI Cooperativa Agropecuária e Industrial, Centro Tecnológico Cocari.

Education: BSc in Agronomy (Universidade Estadual de Maringá, 1991), Plant protection expertise (Universidade Federal de Viçosa, 1997) and MSc in Agronomy (Universidade Estadual de Maringá, 2011).

Address:

BR 376 KM 395 Rural
86975000 Mandaguari, PR Brasil
Phone number: +55 (44) 32330996
Fax number: +55 (44) 32338800
URL: http://www.cocari.com.br
Contact e-mail: ctc@cocari.com.br
alternative e-mail: jbgdias@gmail.com

Research and Professional Experience:

João Batista Gonçalves Dias da Silva is currently coordinator of the Cooperative Technology Center Cocari, Mandaguari PR. Organizes meetings with coffee producers to spread new agricultural techniques and integration between universities, research institutes and producers. Performs planning and conducting experiments with coffee for the supply of material for study in research institutes and universities.

His field of expertise covers the following subjects: plant science, plant protection with emphasis on phytopathology, mainly with major crops, rural outreach events and market development.

Address to access CV: http://lattes.cnpq.br/3498812885030858
Research ID F-7213-2012

Professional Appointments:

Cooperative Technology Center Cocari (2006 - current) - coordinator
Cooperative Technology Center Cocari (2003 - current): technical manager of Seed Laboratory
Autonomous (1994-2001): agronomy engineer

Publications Last 3 Years:

Articles:

Kitzberger, C. S. G., Scholz, M. B. S., Pereira, L. F. P., Vieira, L. G. E., Sera, T., Silva, J. B. G. D., Benassi, M. T. Diterpenes in green and roasted coffee of *Coffea arabica* cultivars growing in the same edapho-climatic conditions. *Journal of Food Composition and Analysis*, v. 30, p. 52-57, 2013 (http://dx.doi.org/10.1016/j.jfca.2013.01.007).

In: Organic Acids
Editor: Cesar Vargas
ISBN: 978-1-63485-931-8

Chapter 4

Effects of Organic Acids and Polysaccharides on the Solubility of Oyster-Derived Zinc Digested *In Vitro*

Yoshiyuki Watanabe*[*]*, Daichi Honda, Yuta Tatewaki, Yoshinori Motonishi and Kazuhiko Yamamoto

Department of Biotechnology and Chemistry, Faculty of Engineering, Kindai University, Takaya, Higashi-Hiroshima, Hiroshima, Japan

Abstract

Zinc is an essential element for the human body as it plays a variety of physiological roles. For example, zinc enhances apoptosis and is necessary to maintain the structure of proteins such as zinc finger proteins. Furthermore, zinc deficiency causes disturbances of growth, taste disorders, and hypogonadism issues. Most dietary zinc is absorbed in the small intestine and its bioavailability depends on the components coexisting in the digested food. The effect of the presence of organic acids and polysaccharides on the solubility of oyster-derived zinc (*Crassostrea gigas*) during *in vitro* digestion was examined using pepsin. The concentration of soluble zinc slightly decreased upon addition of

[*] E-mail address: wysyk@hiro.kindai.ac.jp.

organic acids (e.g., citric, malic, sorbic, or lactic) and remained nearly constant while varying the organic acid to zinc molar ratio. Phytic acid significantly lowered the concentration of soluble zinc, with a negative correlation between the liberated zinc and the added phytic acid. The solubility of zinc during digestion can be affected by organic acid-induced chelation and insolubilization processes. The concentration of soluble zinc slightly decreased in the presence of starch or cellulose. This concentration was independent of the amount of polysaccharide. On the contrary, the concentration of soluble zinc decreased with the added amount of alginic acid, pectin, or chitosan. The content of acidic and basic amino acids (e.g., aspartic acid, glutamic acid, lysine, and histidine) in the oyster protein was *ca.* 30% by mole. Therefore, the electrostatic interaction between an electrolytic polysaccharide and the oyster protein might suppress the solubilization of digested oyster-derived zinc.

Keywords: *In Vitro* digestion, organic acid, oyster, polysaccharide, zinc

1. INTRODUCTION

Zinc is an essential element for the human body as it plays a variety of physiological roles. For example, from 300 to 500 enzymes (e.g., RNA polymerase, alkali phosphatase, and alcohol dehydrogenase) in the human body depends on zinc to exhibit their catalytic activities as this element acts as a co-factor on apoenzymes. Furthermore, zinc enhances apoptosis [1], while being necessary to maintain the structure of proteins such as zinc finger proteins [2] and functioning as an intracellular signaling molecule. Zinc is one of the trace metal elements representing only 0.003% in the human body weight. However, zinc deficiency causes many diseases such as disturbances of growth, taste disorders, and hypogonadism. These diseases can be produced by catalytic, structural, and regulatory malfunctions of zinc. However, most of these mechanisms have remained unclear since the essentiality of zinc in the human body was found. Most of the dietary zinc is absorbed in the small intestine via transcellular pathway. When high concentrations of zinc are present in the lumen, zinc is absorbed via paracellular pathway. Many membrane proteins for transporting zinc have been found in the alimentary canal and the absorption mechanisms have been elucidated.

The bioavailability of a metal element depends on the coexisting components in the digested food. Organic acids are natural compounds and major food components. Complexation of metal ions with organic acids plays an important role in the absorption of minerals by solubilizing these metals under conditions at which they would otherwise be insoluble [3]. Therefore, it is important to build knowledge on the effects of food components (e.g., organic acids) involved in the bioavailability of metals in the gastrointentional conditions. In this chapter, the effects of representative organic acids and polysaccharides on the solubility of zinc derived from oysters (*Crassostrea gigas*) during *in vitro* digestion were examined using pepsin.

2. Materials and Methods

2.1. Materials

Commercial-grade shucked oyster (*Crassostrea gigas*, cultivated in Hiroshima, Japan) was used for *in vitro* digestion. Zinc (purity > 99.999%), pepsin (from porcine stomach mucosa), citric acid, sorbic acid, lactic acid, malic acid, a 50% phytic acid aqueous solution, cellulose, alginic acid, pectin from citrus, chitosan, sodium ascorbate, sodium dodecyl sulfate, 2,2'-bipyridyl, and iron(II) sulfate heptahydrate were purchased from Wako Pure Chemical Industries, Osaka, Japan. Starch and chloral hydrate were obtained from Yoneyama Yakuhin Kogyo, Osaka. Zincon and hydroxylammonium chloride were purchased from Tokyo Chemical Industry, Tokyo, Japan. L-Ascorbic acid was provided by Nacalai Tesque, Inc., Kyoto, Japan. The rest of chemicals were of analytical grade and were purchased from either Wako Pure Chemical Industries or Kishida Chemical Co., Ltd., Osaka.

2.2. Solubilization of Oyster-Derived Zinc during *In Vitro* Digestion

Shucked oysters were homogenized and freeze-dried (DC400, Yamato Scientific Co., Ltd., Tokyo) to obtain powdery sample. The *in vitro* digestion of oyster was carried out by following a modified version of a previously reported procedure [4]. One gram of oyster powder, 100 mg of pepsin, and a specific amount of organic acid or polysaccharide were mixed in 50 mL of a 0.1 mol/L hydrochloric acid solution. Six organic acids (i.e., ascorbic, citric, phytic, sorbic, lactic, and malic) and five polysaccharides (i.e., starch, cellulose, alginic acid, pectin, and chitosan) were used for the *in vitro* digestion of oyster. The pH of the mixture was adjusted to 3.0 by adding a 1 mol/L sodium hydroxide solution. The mixture was incubated at 37°C for 3 h in a water-bath (Personal-11, Taitec Co., Ltd., Saitama, Japan). Twelve milliliters of the mixture were taken out and the sample pH 7.4 was adjusted by adding a 1 mol/L sodium hydroxide solution. After a centrifugation at 6000 rpm for 30 min, 0.2 mL of hydrochloric acid were added to 10 mL of the supernatant. The pH was further adjusted to 7.0 by adding a 6 mol/L sodium hydroxide solution after a filtration.

The concentration of zinc in the digestion sample was measured by the zincon method [5]. 5 mL of the sample were diluted 2 times with distilled water. Then, 0.5 g of sodium ascorbate, 1 mL of a 1% (w/v) potassium cyanide solution, 5 mL of a 0.05 mol/L boric acid buffer solution (pH = 9.0), and 3 mL of a 0.13% (w/v) zincon methanol solution were added to the diluted sample. 3 mL of a 10% (w/v) chloral hydrate solution were added and the sample was stirred for 5 min by a magnetic stirrer at room temperature. The absorbance of the sample at 620 nm was measured by a spectrophotometer (DR4000, HACH Company, Loveland, CO, USA). The calibration curve was obtained by using a high-purity zinc standard. Each measurement was done in triplicate and the mean value was calculated.

2.3. Chelating Ability of Organic Acids and Polysaccharides for Scavenging Metal Cations

The above mentioned six organic acids and five polysaccharides were used at various concentrations for measuring the chelating ability for scavenging the iron(II) cation. The chelating ability was measured by following a previous procedure [6]. First, 0.25 mL of the organic acid or polysaccharide solution and a 0.1 mol/L iron(II) sulfate solution were mixed. Sodium dodecyl sulfate was present in each solution at a concentration of 1% (w/v) in advance. 1 mL of a 0.1 mol/L Tris-HCl buffer solution (pH = 7.4) and 2,2'-bipyridyl, 0.4 mL of a 10% (w/v) hydroxylammonium chloride solution, and 2.5 mL of ethanol were added to the mixture. After appropriate dilution with distilled water, the absorbance at 522 nm was measured using a spectrophotometer. Each measurement was done in duplicate and the mean value was calculated.

2.4. Molecular Weight and Amino Acid Composition of the Oyster Protein

Protein was extracted from shucked oyster by following the method reported elsewhere [7]. Homogenized oyster was added to the equivalent volume of a 0.2 mol/L PBS buffer solution (pH = 7.0). The mixture was stirred at 4°C for 6 h and centrifuged (5000 g) at 4°C for 25 min. Protein was precipitated from the supernatant by adding a 75% (w/v) ammonium sulfate solution at the same volume as the precipitate. The protein sample was dialyzed (6000–8000 Da) with a membrane (Cellu-Sep® T2, Membrane Filtration Products, Inc., Seguin, TX, USA) for 48 h in distilled water and subsequently lyophilized by a freeze-drier.

5 mg of the protein sample were dissolved in 5 mL of a 0.025 mol/L Tris-HCl buffer solution. The molecular weight of the oyster protein was measured using gel filtration chromatography [8]. A glass column (16 mm ϕ × 300 mm) filled with a Sephacryl S-300 gel (GE Healthcare Bio-Sciences AB, Uppsala, Sweden) and distilled water as the eluent was used. The protein solution (1.6 mL) was injected at a flow rate of 5.0 mL/min. The extract was detected with a differential refractometer (YRD-833, Shimamuratech, Tokyo). The calibration curve was prepared using a polyethyleneglycol solution at six different concentrations. The weight-average molecular weights were estimated using calculation software (Chromato-PRO-GPC, Run Tim Co., Kanagawa, Japan).

The protein sample was also used for the amino acid composition measurements. 5 mg of the protein sample were dissolved in 1 mL of a 6 mol/L hydrochloric acid solution and the mixture was incubated at 110°C for 24 h to hydrolyze the protein. After the pretreatment by EZ:faast (Phenomenex, Inc., Torrance, CA, USA), the amounts of free amino acids from the oyster protein were measured by gas chromatography (GC, G-3500, Hitachi, Ltd., Tokyo) with a flame ionization detector. Helium was used as a carrier gas and the flow rate was fixed to 1.8 mL/min. The column temperature was increased at 20°C/min from 80 to 320°C and held at this temperature for 1 min. The injection and detection temperatures were 280 and 320°C, respectively.

3. Results and Discussions

3.1. Effect of the Organic Acids and the Polysaccharides on the Solubility of Zinc from Digested Oyster

Figure 1 shows the relationship between the amount of added organic acid/polysaccharide and the concentration of soluble zinc from oyster after pepsin digestion. The concentration of soluble zinc in the digest with no additives was 4.20 μg/mL. Matsuda et al. reported a solubility of zinc of 93.3% (pepsin digest, pH = 3.0) [4]. Since the zinc content in the used oyster was 123 μg/g, the amount of zinc eluted from 1 g of oyster sample was calculated to be 114 μg (versus 92.4 μg as estimated herein). This seems to be mainly produced by an inhomogeneous zinc content distribution in the oyster. The concentration of zinc with ascorbic acid was nearly similar as the control. The addition of citric, malic, sorbic, or lactic acids slightly decreased the concentration of zinc to the same levels for all the organic acids. Similar results have been reported on zinc solubilization from fortified cereals [3]. In addition, the concentration remained nearly unchanged with the organic acid to zinc molar ratio. Phytic acid significantly lowered the concentration of soluble zinc, and a negative correlation between the liberated zinc and the added phytic acid was observed. The chemical properties of the organic acids were shown to contribute to the solubility of zinc from oyster. Similarly to ascorbic acid, neutral polysaccharides such as starch and cellulose did not affect the solubility of zinc. On the contrary, acidic or basic polysaccharides such as alginic acid, pectin, and chitosan lowered the zinc concentration in the digest, particularly at high amounts of added polysaccharide. These tendencies can be potentially ascribed to the electrolytic dissociation of the polysaccharides.

3.2. Chelating Ability of the Organic Acids and the Polysaccharides in Zinc Solution

The chelating activities of the organic acids and the polysaccharides used for the *in vitro* digestion of oyster were evaluated to examine their effect on the elution property of zinc in the digest. Figure 2 shows the chelating activity of the organic acids for scavenging the iron(II) cation. The Iron(II) cation can form a complex with the 2,2'-bipyridyl molecule. The presence of chelation agents inhibited the complex formation, resulting in lower absorbance values for the sample. The chelating activity was found to depend on the organic acid in the following order: phytic acid > citric acid > malic acid > ascorbic acid, lactic acid > sorbic acid. Unlike phytic acid, sorbic, lactic, and ascorbic acids hardly exhibited chelating activity. Phytic acid has been known to form a strong and insoluble complex with zinc [9]. Therefore, the low concentration of zinc in the digest in the presence of phytic acid (Figure 1) can be attributed to the insolubilization of zinc upon chelation with phytic acid. However, despite their different chelating activities, similar zinc solubilities were obtained for the rest of organic acids regardless the acid concentration. These results might be explained in terms of the different solubility of the chelate complexes. Polysaccharides did not chelate the iron(II) cation at the range of concentrations used for the *in vitro* digestion (data not shown). It is suggested that the elution behavior of zinc in the digest containing additives such as organic acids and polysaccharide cannot be explained exclusively by chelation.

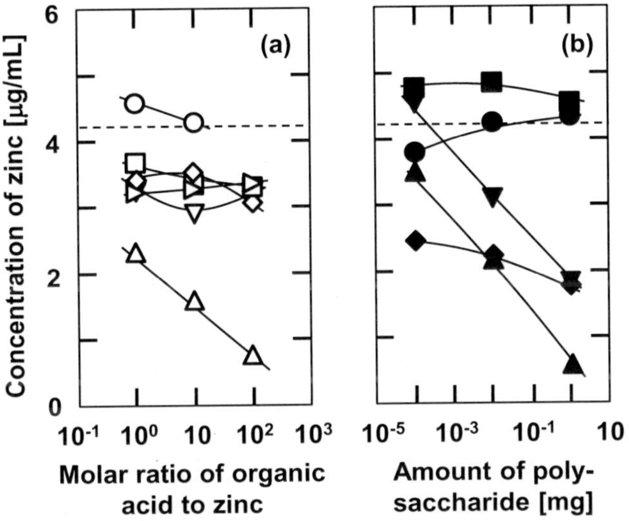

Figure 1. Effects of: (a) organic acids; (○) ascorbic acid, (□) citric acid, (△) phytic acid, (◇) sorbic acid, (▽) lactic acid, and (▷) malic acid, and (b) polysaccharides; (●) starch, (■) cellulose, (▲) alginic acid, (◆) pectin, and (▼) chitosan on the solubilization of zinc from oyster under pepsin digestion at 37 °C and pH = 3.0. The broken curves represent the concentration of zinc in the absence of organic acid or polysaccharide.

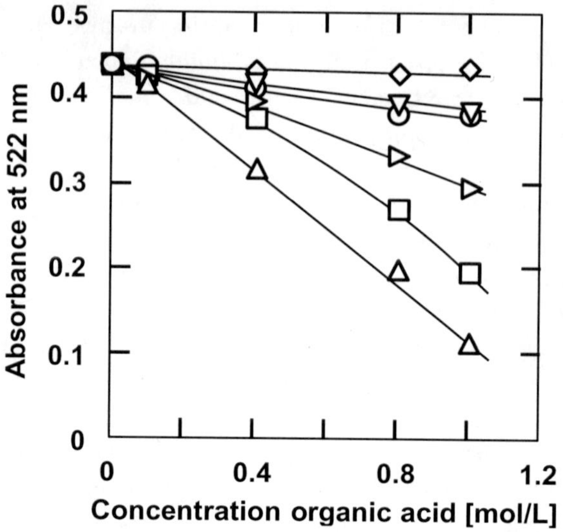

Figure 2. Iron(II) cation chelating activity of the organic acids. Similar symbols as in Figure 1 (a).

3.3. Relationship between the Elution of Zinc from the Digested Oyster and the Chemical Structure of the Oyster Protein

In order to investigate the effect of the oyster protein on the solubility of zinc during the *in vitro* digestion, the amino acid composition of this protein was measured. The protein content was 0.40 g/g, with an average molecular weight estimated to be 240000. Table 1 shows the amino acid molar composition of the oyster proteins. The total content of acidic and basic amino acids (e.g., aspartic acid, glutamic acid, lysine, and histidine) in oyster protein was high (*ca.* 30% by mole). Furthermore, the contents of highly polar threonine and tyrosine bearing a hydroxyl group in the residue were also high. Some of these amino acids have been reported to enhance the solubility of zinc under gastrointentional conditions [3]. Thus, the protein and its hydrolysate would also directly and/or indirectly have an effect on the elution of zinc via digestion by a protein hydrolase. The electrostatic interaction between an electrolytic polysaccharide and the oyster protein might form macro- and inter-molecular networks allowing the sealing of zinc in the network while suppressing its solubilization from the digested oyster.

Table 1. Amino acid molar composition in oyster proteins (n = 3)

Amino acid	Molar composition [%]
Alanine	7.76 ± 1.64
Glyscine	7.19 ± 0.44
Valine	1.35 ± 0.09
Leucine	14.91 ± 0.33
Isoleucine	3.76 ± 0.22
Threonine	6.57 ± 0.57
Serine	2.53 ± 0.20
Proline	11.39 ± 1.01
Asparagine	0.39 ± 0.40
Thiaproline	0.35 ± 0.25
Aspartic acid	9.06 ± 1.05
Methionine	0.65 ± 0.40
Hydroproline	0.58 ± 0.24
Glutamic acid	5.34 ± 0.67
Phenylalanine	4.30 ± 0.24
α-Aminoadipic acid	0.26 ± 0.16
α-Aminopimelic acid	0.32 ± 0.12
Glutamine	n.d.
Ornithine	0.16 ± 0.04
Lysine	11.12 ± 0.61
Histidine	4.93 ± 0.80
Tyrosine	4.64 ± 0.47
Cystathionine	1.21 ± 0.52
Cystine	1.23 ± 0.63

CONCLUSION

The solubility of zinc from oysters in the presence of other food components such as organic acids or polysaccharides during *in vitro* digestion should be examined from multiple viewpoints because of the complexity of the system. With this aim, some devices and procedures are required to study the interaction among the various chemical species present in the mixture.

REFERENCES

[1] Chimientia, F.; Sevea, M.; Richardb, S.; Mathieub, J.; Faviera, A. Role of Cellular Zinc in Programmed Cell Death: Temporal Relationship between Zinc Depletion, Activation of Caspases, and Cleavage of Sp Family Transcription Factors1, *Biochemical Pharmacology* 2001, 62 (1), 51-62.

[2] Ganss, B.; Jheon, A. Zinc Finger Transcription Factors in Skeletal Development, *Critical Reviews in Oral Biology and Medicine* 2004, 15 (5), 282-297.

[3] Desrosiers, T.; Clydesdale, F. M. Effectiveness of Organic Chelators in Solubilizing Calcium and Zinc in Fortified Cereals under Simulated Gastrointestinal pH Conditions, *Journal of Food Processing and Preservation* 1989, 13, 307-319.

[4] Matsuda, Y.; Sumida, N.; Yoshida, M. Zinc in Oysters (Crassostrea gigas): Chemical Characteristics and Action during in vitro Digestion, *Journal of Nutritional Science and Vitaminology* 2003, 49 (6), 405-408.

[5] Rush, R. M.; Yoe, J. H. Colorimetric Determination of Zinc and Copper with 2-Carboxy-2-hydroxy-5-sulfoformazylbenzene, Analytical Chemistry 1954, 26 (8), 1345-1347.

[6] Yamaguchi, F.; Ariga, T.; Yoshimura, Y.; Nakazawa, H. Antioxidative and Anti-glycation Activity of Garcinol from *Garcinia indica* Fruit Rind, *Journal of Agricultural and Food Chemistry* 2000, 48, 180-185.

[7] Wanga, J.; Huc, J.; Cuid, J.; Baia, X.; Dua, Y.; Miyaguchie, Y.; Lin, B. Purification and Identification of a ACE Inhibitory Peptide from Oyster Proteins Hydrolysate and the Antihypertensive Effect of Hydrolysate in Spontaneously Hypertensive Rats, *Food Chemistry* 2008, 111, 302-308.

[8] Watanabe, Y.; Morishita, T.; Tojo, H.; Toriyama, E.; Watariue, N.; Yamaguchi, R.; Nomura, M. *In Advances in Chemistry Research 27*; Nova Science Publishers, Inc.: New York, 2015; pp. 19-32.

[9] Lönnerdal, B. Dietary Factors Influencing Zinc Absorption, *Journal of Nutrition* 2000, 130 (5), 1378S-1383S.

In: Organic Acids
Editor: Cesar Vargas
ISBN: 978-1-63485-931-8

Chapter 5

EVAPORATION OF ORGANIC ACIDS AQUEOUS SOLUTIONS THROUGH SPREAD FILMS OF POLYELECTROLYTE/SURFACTANT COMPLEXES

V. Kuznetsov**, *A. Akentiev and V. Rakhimov
Institute of Chemistry of the St. Petersburg State University,
St. Petersburg, Russian Federation

ABSTRACT

Wilhelmy plate technique was used to determine surface tension isotherms of aqueous solutions of formic and acetic acids with concentration up to 30 vol% and with a film of solution of polyelectrolyte/surfactant complexes spread to the surface. Complexes of sodium polystyrene sulfonate / dodecyl trimethyl ammonium bromide and poly (diallyldimethylammonium chloride)/ sodium dodecylsulfate were used. The formation conditions of films with the extreme concentration of the complexes were determined. The films reduce the evaporation rate of acids solutions by 3-6% and demonstrate selective properties increasing the water content in the vapor by 1-5 abs%. Evaporation tests carried out by means of helium blowing over surface of the liquid with spread film.

* kuvik2007@yandex.ru.

Keywords: Organic acids, polyelectrolyte/surfactant complexes; spread films; surface tension; rate of evaporation; selectivity

INTRODUCTION

Presence of surfactants in homogeneous and heterogeneous systems affects the evaporation rate of components, usually reducing it. Beverley et al. [1] used a gravimetric technique to show that evaporation of aqueous solutions of nonionic surfactants (n-dodecyl hexaoxyethylene glycol ether) is slowing down due to concentration gradients in the fluid. Diffusion resistance to mass transfer is determined by the liquid phase. Some surfactants reduce the evaporation rate of slightly superheated water at 100°C owing to a decrease of convective surface tension at the interface [2]. As the concentration of a surfactant increases, the influence of its chemical nature on the evaporation reduces. According to Francis and Berg [3], adding a surfactant to a binary mixture increases the efficiency of a packed distillation column. 1-decanol was added in the separation of an aqueous solution of formic acid. When testing organic compounds as surface-active additives the authors used silicone or fluorosilicone oils as well as alkyl polyoxyethylene glycol. They explained the improvement of the efficiency of separation by the stabilization of the fluid film overlying the nozzle with an increase of the mass transfer surface. Data on the mass transfer of mesitylene dissolved in water as air bubbles in the presence of an ionic surfactant showed that the surfactant does not affect the effectiveness of evaporation until surface bubbles coverage does not exceed 0.7 [4]. The authors noticed a certain reduction in the mass transfer coefficient in the case of the interaction between the hydrocarbon and the surfactant molecules. The authors used the method of inverse gas chromatography to study the evaporation rate of ethanol and trichloroethylene into nitrogen at 306.2 K in the presence of the Triton X-100 surfactant [5]. They noted that a substantial reduction of the evaporation rate requires a significant amount of surfactant necessary for the formation of densely packed surface monolayers, including insoluble ones. Shen et al. [6] describe a study of evaporation of benzene, toluene, and ethyl benzene dissolved in water. They examined the effect of added surfactants: sodium dodecyl benzene sulfonate (SDBS), cetyl trimethyl ammonium bromide (CTAB) and polyoxyethylene (4) lauryl ether (Brij30). The results show that if the concentration of a surfactant exceeds the critical micellization concentration, evaporation rates of organic additives decrease. The authors felt that the main reason for this is the formation of micelles, which leads to lower concentrations of organic additives in the interface as compared to the bulk. Benzene, toluene, and ethyl benzene vaporized worse when using CTAB and Brij30 as compared to SDBS. The authors found a positive correlation between an inhibitory ability of surfactants and Henry's constants of organic substances.

Thus, dissolved surfactants reduce the evaporation at sufficiently high concentrations, which affects the picture of motion and mass transfer in the liquid as well as heat transfer including those at the interface. In addition, they are able to increase the surface of mass transfer.

Insoluble surfactant spread monolayers at the water surface display the reduction of the evaporation rate, which is almost absent for adsorption films [7-10]. Study of this phenomenon for the water/air system started decades ago [7, 11] and was interpreted as due to the reduction of the evaporation area owing to the presence of surfactant molecules at the surface of the liquid [12] or the probability of formation of "holes" in monolayers comparable in size with water molecules [13, 14]. However, in these assumptions, it is difficult to explain the difference between the influences on evaporation caused by spread and adsorbed films. Apparently, more physically motivated model is that of Barnes [8]. Under this model, insoluble monolayers at the water surface are two-dimensional crystalline microdomains surrounded by relatively loose structures through which evaporation occurs predominantly.

In [15] we showed that spread films of polyelectrolyte/surfactant complexes also reduce water evaporation. In particular, the studied complexes were sodium polystyrene sulfonate/dodecyl trimethyl ammonium bromide (PSS/DTAB) and poly(diallyldimethylammonium chloride) / sodium dodecyl sulfate (PDADMAC/SDS). We can assume that, in the case of liquids containing other volatile substances in addition to water, the evaporation rate will change. This will change the composition of the vapor.

In [4, 16-21] the authors examine evaporation in systems containing other components besides water. These studies focus on the evaporation of surfactant-stabilized water emulsions and solutions of hydrocarbons in water, as well as mixtures of hydrocarbons having surfaces covered with a film of a surfactant dissolved in water. Surface properties of such systems in the presence of surfactants, as well as evaporation of their components are not subject to detailed analysis.

Interaction between oppositely charged polymers and surfactants in aqueous solutions acquires a growing interest [22-25]. Such interactions often lead to the formation of complexes in the bulk of a solution; thus, their size and properties can vary [26]. Even at low concentrations, complexes are able to adsorb on the surface of the water-air interface; monolayers can also form depending on the composition. The driving forces of such complex formation are electrostatic and hydrophobic interactions between polyelectrolytes and surfactants. Adsorption films of the complexes can vary significantly with surface tension, but one believes [27] that their thickness is small and polymeric chains are stretched across the surface. Mixed polymer/surfactant layers are metastable systems [28]. Because of adsorption, they behave as insoluble films and their structure depends on the adsorption technique.

Along with the studies of polyelectrolyte/surfactant complex films on the water/air interface, recently there appeared studies of such films on the water/oil interface [29] and in systems with liquid crystals [30].

However, there are no studies on the evaporation of liquids with spread films of polyelectrolyte/surfactant complexes that contain components other than water. The aim of this work is an experimental investigation of evaporation of organic acids from aqueous solutions through spread films. The paper presents the results of surface tension measurements in aqueous solutions of organic acids with spread films of PSS/DTAB or PDADMAC/SDS solutions together with the impact of the films on evaporation and composition of the vapor. The data obtained will be useful for better understanding the role of spread films of surfactant complexes in mass transfer in the evaporation and desorption processes.

EXPERIMENTAL

The surface tension was measured by the Wilhelmy plate method by means of the Sigma Force Tensiometer 700 (Finland) with a platinum plate. PTFE measuring cuvette was 40 cm long and 11 cm wide. Evaporation rate of the liquid and vapor composition were determined using the unit described in [15]. A required amount of the solution of the PSS/DTAB or PDADMAC/SDS complex was spread on the surface of the aqueous solution of organic acids with a microsyringe; the generated vapor was removed with helium of zero humidity. The temperature of the solutions was 22 ± 0.1 °C. The optimum speed of helium over the surface of the liquid was 1.2 cm s^{-1}. The vapor was collected in a condensation trap cooled with liquid nitrogen. Its weight gain together with the time of the experiment allowed the calculation of the evaporation rate of the liquid. Vapor composition was analyzed by gas chromatography (GC-2010PLUS device, Shimadzu, Japan) of the trap content. The content in the vapor is small for solutions of acids with low concentration, thus the results of the chromatographic analysis are unreliable. For this reason, for the study we chose solutions with acid concentrations 20-30 vol%.

We used the deionized water to prepare a solution of a PSS/DTAB or a PDADMAC/SDS complex and in the experiments of the study of surface properties and evaporation of solutions. Sodium polystyrene sulfonate and poly(diallyldimethylammonium chloride) (Sigma-Aldrich) were used without further purification. Dodecyl trimethyl ammonium bromide and sodium dodecyl sulfate (Sigma-Aldrich) were purified by recrystallization from a mixture of ethyl acetate and ethanol and pure ethanol consequently. Formic and acetic acids were used after rectification.

Solutions for spreading polyelectrolyte/surfactant films on the water - organic acid surface were prepared in such a way that a monomeric unit of the polyelectrolyte had one molecule of a low molecular weight surfactant. This ratio of molecules of the complex on the surface of the liquid results in the formation of the most stable film without an excess of any component [31]. Changing the concentration of the complex on the surface of a binary liquid was carried out with the microsyringe by applying drops of the solution on the surface.

RESULT AND DISCUSSION

Increasing the content of formic and acetic acids in water sharply reduces its surface tension. However, as follows from Figures 1-4, application of polyelectrolyte/surfactant complexes leads to a further decrease in the surface tension. Yet, when acid concentrations in the solution increase, the decrease in surface tension occurs in the shrinking range reaching a minimum of 37-40 mN m^{-1}. Further addition of the complex does not lead to a noticeable change of the surface tension, and one can believe that the excess complex was dissolved. When the content of acids in the solution increases 30 vol%, the surface tension without spread film takes such a low value that the addition of the complex almost does not affect the surface tension, therefore the data are not represented in the figures. For relatively concentrated solutions of the acids after applying the films of the complex, the surface tension isotherms almost coincide (Figure 2 and 4). Apparently, in this case, the film is saturated with the acid and its properties slightly change after concentrating the solution. Within a few hours after spreading the film, in the studied surface concentration range the surface tension does not change, thus indicating the film stability.

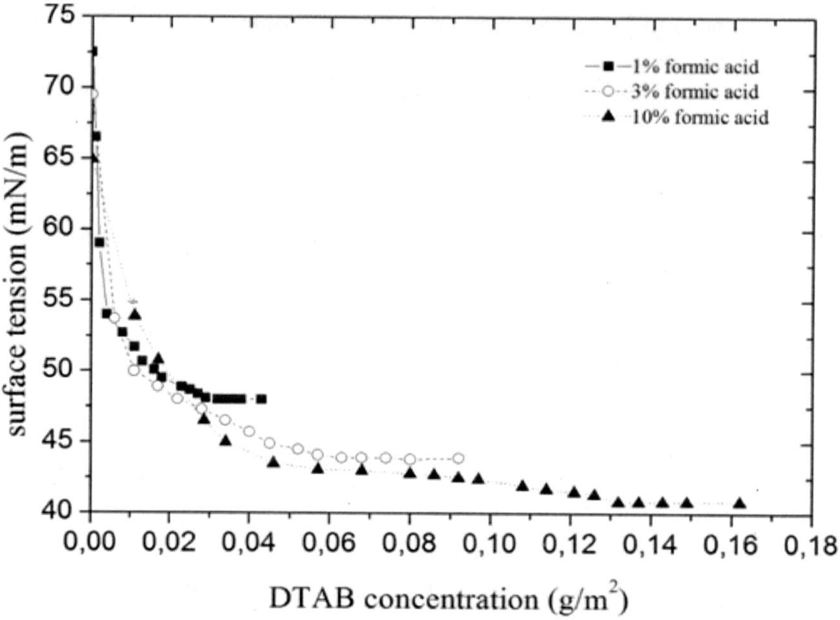

Figure 1. Surface tension isotherm for the solution of the PSS/DTAB complex spread to solutions of formic acid (low acid concentrations).

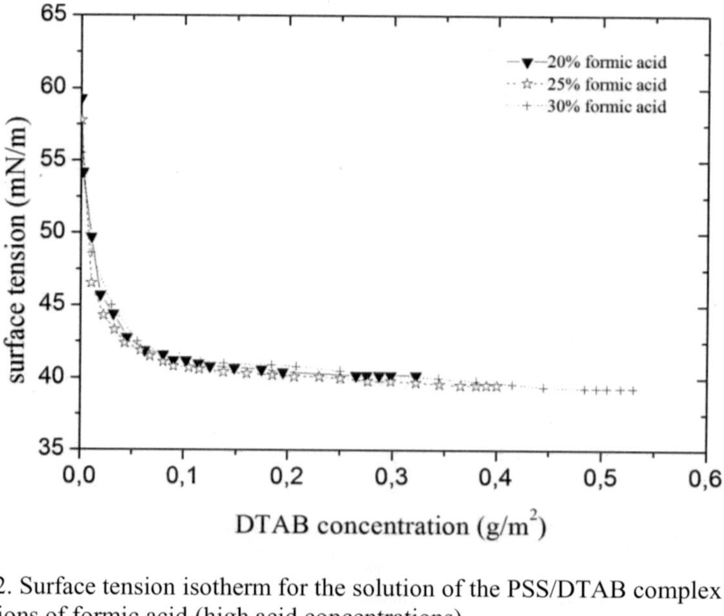

Figure 2. Surface tension isotherm for the solution of the PSS/DTAB complex spread to solutions of formic acid (high acid concentrations).

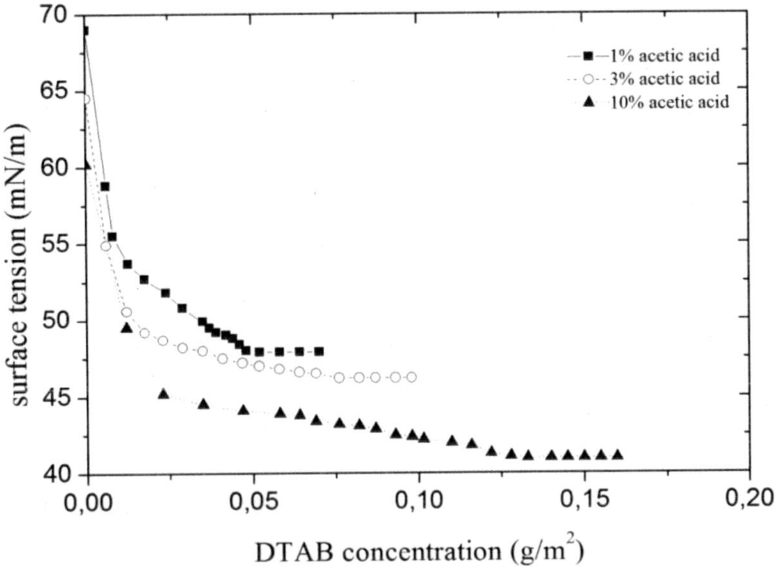

Figure 3. Surface tension isotherm for the solution of the PSS/DTAB complex spread to solutions of acetic acid (low acid concentrations).

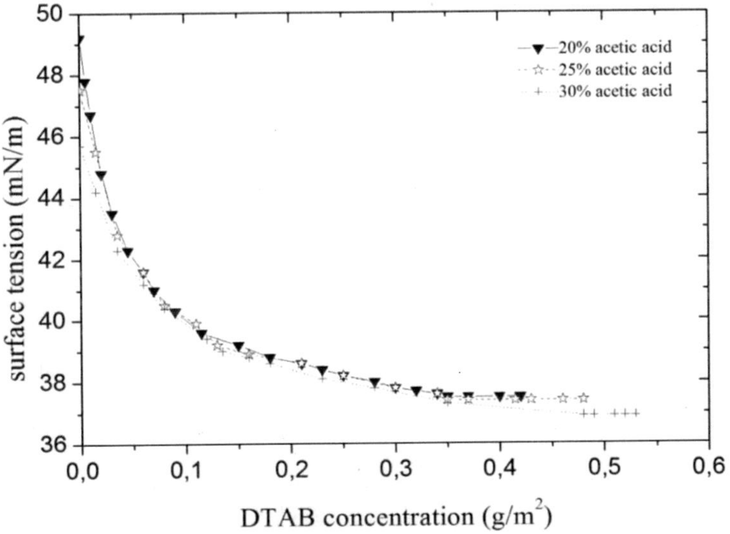

Figure 4. Surface tension isotherm for the solution of the PSS/DTAB complex spread to solutions of acetic acid (high acid concentrations).

The surface tension data made it possible to identify the amount of the polyelectrolyte/surfactant solution, which produced on the surface of the aqueous solution of organic acids a film with extreme concentration of the PSS/DTAB or PDADMAC/SDS complex, namely the surface concentration when the surface tension of the solution acquires a constant value. Figures 5 and 6 demonstrate the dependences of the vapor flow on the liquid concentration for a clean surface (dashed lines) and for the liquid with spread PSS/DTAB films (dotted lines). As you can see from the plots, evaporation rate of acetic acid solutions exceeds this value for formic acid solutions. We explain it by the difference of molecular weights of the acids. In addition, the presence of the film of the complex reduces the rate of evaporation of the liquid by 3-6%. In [31] we used atomic force microscopy (AFM) to study films of the PSS/DTAB complex transferred from the water surface to the substrate of mica. At concentrations of polyelectrolyte and surfactant used in [31] the values of the surface tension of the film transferred on the mica surface was about 40 mN/m. This value is close to the concentrations of PSS and DTAB for which we conducted the evaporation study. AFM images show that thin segments of the films quite uniformly alternate with segments several nanometers high. We can assume that the film formed on the surface of a binary liquid has a similar structure, and there are sites through which evaporation occurs.

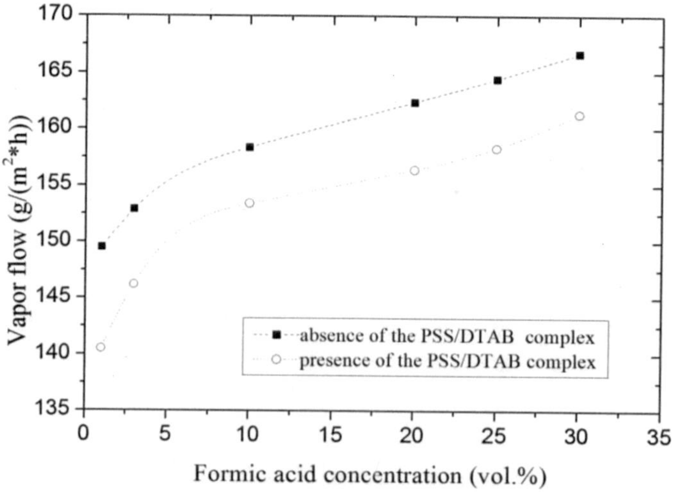

Figure 5. Dependence of the vapor flow on the concentration of formic acid in the solution.

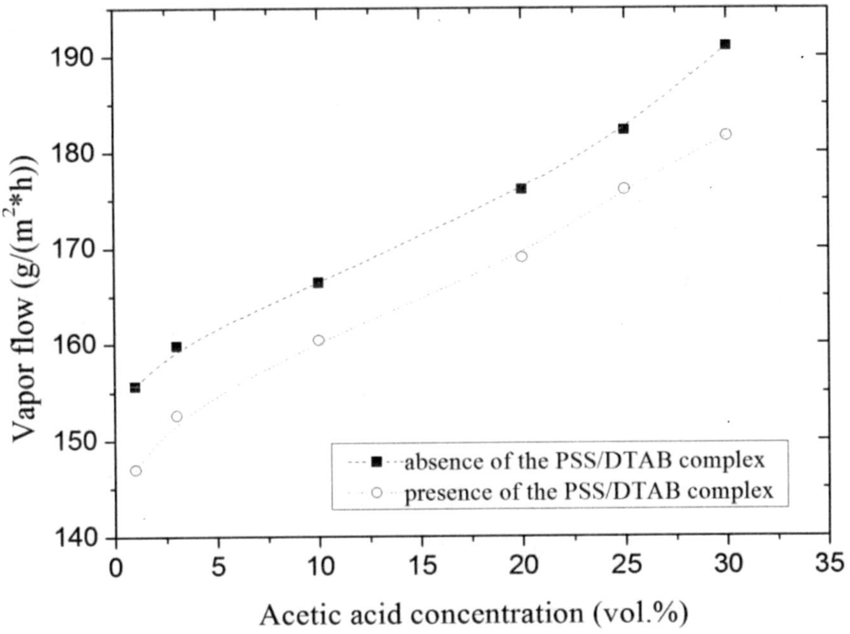

Figure 6. Dependence of the vapor flow on the concentration of acetic acid in the solution.

The chromatographic analysis of the vapor presented in Figures 7 and 8 shows an increased content of acetic acid in vapors compared to formic acid. The presence of the monolayer of the PSS/DTAB complex on the surface of aqueous solutions of acids results in a small (2-5 abs%) reduction of acid content in the vapor, i.e., the emergence of selectivity of the film. The reason for this could be the following. Hydrophilic constituents of the polyelectrolyte/surfactant complex located on the surface of the liquid direct into the bulk of the solution. This leads to an increase in the concentration of water in the boundary layer of the liquid and the reduction of the evaporated acid flow with a corresponding change in the composition of the vapor. There are transport restrictions because of the difference in molecular weights and sizes of organic acids and water, as well as their interaction with the molecules of the complex in diffusion through the film. Clarification of reasons for selectivity requires further theoretical and experimental research.

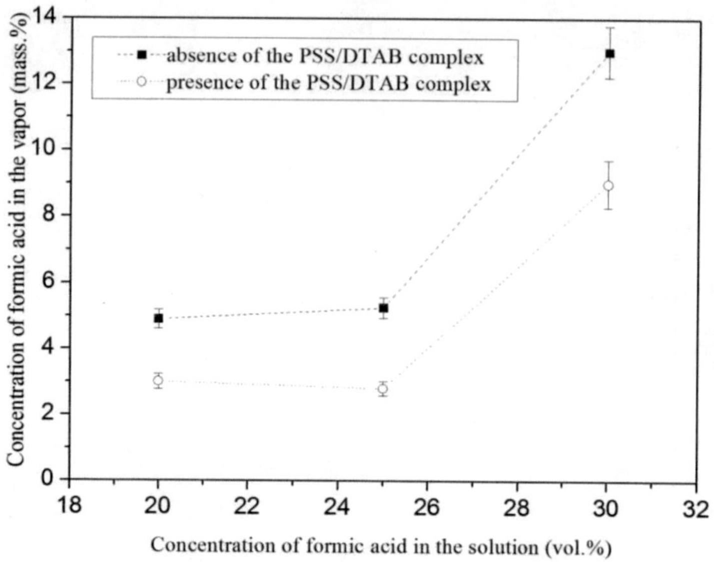

Figure 7. Dependence of the concentration of formic acid in the vapor on the concentration of formic acid in the solution.

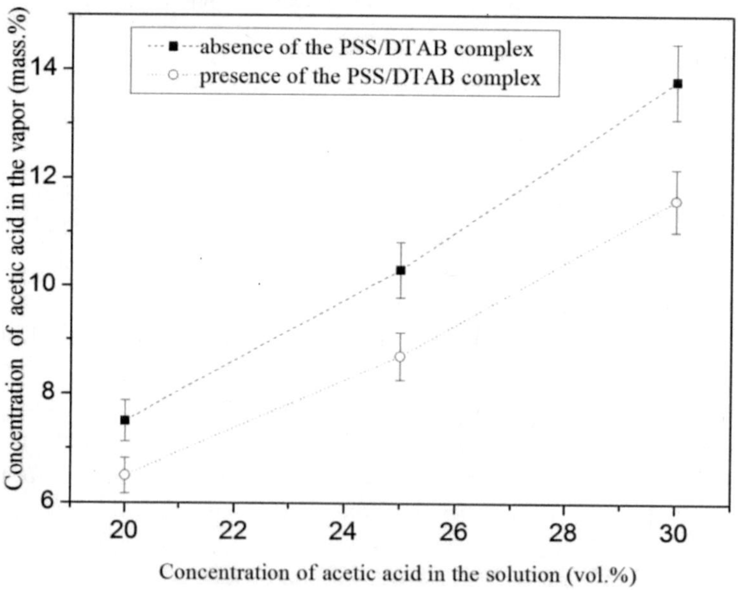

Figure 8. Dependence of the concentration of acetic acid in the vapor on the concentration of acetic acid in the solution.

The results are consistent with a model of polyelectrolyte/surfactants films, according to which we assume hydrophobic interactions between hydrocarbon tails of surfactant molecules, between the tails and hydrophobic parts of polyelectrolyte segments, as well as between the segments [32]. These interactions lead to the formation of aggregates close to two-dimensional, which impede evaporation.

The PDADMAC/SDS complex demonstrates similar surface properties and influences the evaporation rate and vapor composition, so we don't present the experimental data in this paper.

CONCLUSION

Selectivity of binary liquid separation by evaporation depends on the ratio of relative volatility if the process conditions are close to equilibrium. In pervaporation, when a membrane, commonly polymeric, separates a liquid and a vapor, the separation efficiency is determined, in addition to the characteristics of the liquid-vapor phase transition, by physico-chemical properties of the polymer. The spread film of polyelectrolyte/surfactant complex is also a membrane. On the surface of water solutions of formic and acetic acids with acid concentration up to 30 vol% the film is stable and reduces the surface tension. Evaporation of solutions in the presence of spread films with extreme concentration of the complex is characterized by a decrease in the rate and low acid content. Confirmation of this effect is the goal of further experimental studies. They will take into consideration other complexes of synthetic polyelectrolytes and low molecular weight surfactants together with aqueous solutions of organic solvents.

REFERENCES

[1] Beverley, K.J.; Clint, J.H.; Fletcher, P.D.I. *Phys. Chem. Chem. Phys.* 2000, 2, 4173-4177.

[2] White, I. *Ind. Eng. Chem. Fundam.* 1976, 15, 53-59.

[3] Francis, R. C.; Berg J.C. *Chem. Eng. Sci.* 1967, 22, 685-692.

[4] Medrzycka, K.B. *Sep. Sci. Technol.* 1992, 27, 1077-1092.

[5] Gavril, D.; Atta, K.R.; Karaiskakis, G. *AIChE J.* 2006, 52, 2381-2390.

[6] Shen, X.-Y.; Sun, J.-J.; Ma, Z.-Y.; Luo, X.-L. Huanjing Kexue/*Environmental Sci.* 2005, 26, 122-126.

[7] Barnes, G.T. *Adv. Colloid Interface Sci.* 1986, 25, 89-200.

[8] Barnes, G.T. *Colloids Surf.* A. 1997, 26, 149-158.

[9] Lunkenheimer, K.; Zembala, M. J. *J. Colloid Interface Sci.* 1997, 188, 363-371.

[10] Fainerman, B.; Makievski, A.; Kragel, J.; Javadi, A.; Miller, R. *J. Colloid Interface Sci.*, 2007, 308, 249-253.

[11] Retardation of Evaporation by Monolayers: *Transport Processes*; Ed. by La Mer, K.; Academic Press: N Y, 1962.

[12] Barnes, G.T.; Quickenden, T.I.; Saylor, J.E. *J. Colloid Interface Sci.* 1970, 33, 236-243.

[13] Blank, M. J. Phys. Chem. 1964, 68, 2793-2800.

[14] Blank, M.; Britten, J.S. J. Colloid Sci. 1965, 20, 789-800.

[15] Kuznetsov, V.M.; Akentiev, A.V.; Noskov B.A.; Toikka A.M. *Colloid J.* 2009, 71(2):202-207.

[16] Friberg, S.E. J. Dispersion Sci. Technol. 2006, 27, 573-577.

[17] Aranberri, I.; Binks, B.; Clint, J.H.; Fletcher, P.D.I. *Langmuir* 2004, **20,** 2069-2074.

[18] Rusdi, M.; Moroi, Y.; Nakahara, H.; Shibata, O. *Langmuir* 2005, 21, 7308-7310.

[19] Taflin, D. C.; Zkhang, S. H.; Allen, T.D.; Davis, E. J. *AIChE J.* 1988, 34, 1310-1320.

[20] Shen, X.-Y.; Sun, J.-J.; Ma, Z.-Y.; Luo, X.-L. Huanjing Kexue/*Environmental Sci.* 2005, 26, 122-126.

[21] Leonard, J.T. Suppression of evaporation of hydrocarbon liquids and fuels by films containing aqueous film forming foam (AFFF) concentrate FC-196. Rep. *NRL Prog.* January 1975: 23.

[22] Piculell, L.; Lindman, B. *Adv. Colloid Interface Sci.* 1992, 41(C), 149-178.

[23] Goddard E.D.; Ananthapadmanabhan K.P. *Interaction of Surfactants with Polymers and Proteins*; CDR Press: Boca Raton, FL, 1993.

[24] Kwak J.C.T. Polymer-surfactant systems; *Surfactant Sci.* Series; 1998, 77.

[25] Claesson P.M.; Dedinaite A.; Poptoshev E. *Physical Chemistry of Polyelectrolytes*, Ed. Radeva T.; Marcell Deccer: NY, 2001; 99, 447-473.

[26] Bain C.D.; Claesson P.M.; Langevin D.; Meszaros R.; Nylander T.; Titmuss S.; von Klitzing R. *Adv. Colloid Interface Sci.* 2010, 155(1-2), 32-49.

[27] Langevin D. Adv. Colloid Interface Sci. 2001, 89-90, 467-484.

[28] Langevin D.; Monroy F. *Curr. Opinion Colloid Interface Sci.* 2010, 15(4), 283-293.

[29] Aidarova S.; Sharipova A.; Kragel J.; Miller R. *Adv. Colloid Interface Sci.* 2014, 205, 87-93.

[30] Lockwood N.A.; Gupta J.K.; Abbott N.L. *Surface Sci. Rep.* 2008, 63(6), 255-293.

[31] Akentiev A.; Bilibin A.Yu.; Zorin I.M.; Lin S.Y.; Loglio G.; Miller R.; Noskov B.A. *Colloid J.* 2011, 73(4), 437-444.

[32] Noskov B.A.; Loglio G.; Miller R.J. *J. Phys. Chem.* B 2004, 108(48), 18615-18622.

In: Organic Acids
Editor: Cesar Vargas
ISBN: 978-1-63485-931-8

Chapter 6

PERVAPORATION OF ACETIC ACID AQUEOUS SOLUTION: INFLUENCE OF LIQUID SORPTION AND EFFECT OF DOWNSTREAM PRESSURE ON SEPARATION PERFORMANCE

V. Kuznetsov* and A. Pulyalina
Institute of Chemistry of the St. Petersburg State University,
St. Petersburg, Russian Federation

ABSTRACT

Pervaporation is a more energy saving, *environmentally* safe and *clean* technology of liquid mixture separation as compared with the existing techniques such as distillation. At present, the effective separation of aqueous organic solutions by using pervaporation is one of the actual tasks of membrane technology. The effectiveness of liquid separation considerably depends on the conditions of the experiments and sorption properties respect to liquid mixtures.

A manuscript presented the influence of sorption capacity and downstream pressure on pervaporation results. Acetic acid aqueous solution were used as model separation mixtures. The main sorption characteristics (equilibrium swelling and composition of sorbates) with respect to various compositions of feed mixtures were studied. Pervaporation experiments were performed in the range of 1 - 50 mm Hg

* kuvik2007@yandex.ru.

of downstream pressure using the membranes based on cellulose hydrate. It was shown that membranes exhibit higher dehydrating properties with increase water concentration in vapor. The opposite trend - decrease of pervaporation flux and increase of separation factor with the increases of downstream pressure was observed. It was found that contribution of swelled membrane layer to the values of selectivity was significant and achieved up to 60% at downstream pressure of 30 mm Hg and higher.

Keywords: organic acids, sorption, swelling, downstream pressure, pervaporation

INTRODUCTION

Pervaporation (PV) has been widely used in separation technology when traditional methods, such as distillation or rectification, are not applicable [1-2]. The main advantages of PV is ability of azeotropic and close-boiling point mixtures separation [3-5]. Application area of PV is dehydration of organic solvents, evaporation of volatile organic compounds from aqueous solutions, and separation of mixed anhydrous organic mixtures [2]. At present, the generally accepted mechanism of mass transport through dense polymeric membranes is based on solution-diffusion theory [6-7]. Thus, the separation occurs because of the different rates of sorption and diffusion of the feed components through the membrane [8-11].

By now, a number of methods for simulation and calculation of pervaporation parameters have been developed basically using solution-diffusion theory [1, 11-14]. At the same time, the concepts of the mass transport in PV are widely various and sometimes contradictory [15]. Therefore, the existing methods are insufficient as comparison vapor-liquid equilibrium in the case of distillation [16]. This fact very limited industrial application of PV.

Development of pervaporation theory requires additional physicochemical investigation of stages of the separation process. Study of sorption as an equilibrium between a feed mixture and a membrane was the subject of extensive research [17]. In [18] the equilibrium between ternary mixture of benzene, cyclohexene, and cyclohexane and films of polyurethane and polyetherblockamide was studied. In [19] study of sorption of aqueous solutions of ethanol, methanol, and 2-propanol using polyvinyl alcohol film was reported. The sorption of gaseous oxygen, carbon dioxide, ethylene, dimethyl sulfide, trichloroethylene, and toluene by polydimethylsiloxane film was examined in [20]. These data was used for component sorption selectivity calculation defined as the ratio of the component concentration in the membrane to the concentration in separation solution. This factor made it possible to prognosticate the pervaporation selectivity of the membrane material [18].

At the vacuum condition, pervaporation is an open phase process [21] because of permeate is continuously removed. Evaporation through the membrane has some similarities with evaporation when phase compositions, partial pressure of components correspond to vapor-liquid equilibrium. But pervaporation separation was carried out using membrane separated liquid and vapor phase. As the residual pressure increases, the flux of substances across the membrane decreases and the permeate composition changes. It is noteworthy that the dependence of pervaporation parameters on the residual pressure has been the subject of only a small number of reports. Usually only a low permeate pressure is necessary for producing technologically significant high fluxes [22]. In [23], the concept of a vapor–liquid pseudo-equilibrium in pervaporation was regarded at a zero flux and different pressures on the sides of feed and vapor.

Acetic acid is an important raw material in industry and has been widely used as solvent in food, pharmaceutical, chemical, and dyes industries. At present a large number of wastewaters containing acetic acid at different concentrations emerged from those industries, including the production of acrylic acid, cellulose acetate, terephthalic acid, poly(vinyl alcohol), and acetaldehyde by the Wacker process, destructive distillation of wood, and reactions involving acetic anhydride, etc. [24]. Separation of acetic acid from dilute aqueous solutions is very difficult, even implausible because of the close boiling point of two compounds. To improve the efficiency of separation of acetic acid from water solution and reduce energy consumption pervaporation can be used.

Some examples of sorption study for explanation of mass transfer in pervaporation of water–acetic acid mixtures are presented in [25-28].

An experimental evaluation using polyphenylsulphone membranes to separate the acetic acid–water mixture is performed in [25]. The detailed study on membrane swelling showed that the increase of degree of swelling is mainly caused by the molecular mass of acetic acid. Due to the importance of the membrane top layer in PV separation, its surface sorption was studied and analyzed using Fourier transform infrared spectroscopy to investigate preferential sorption on the top layer of the membranes. Data indicate that the sorption of acetic acid increases with an increased acetic acid in the feed composition and that water interaction is better at low water content in the membrane.

Novel polyelectrolytes complex (PEC)/11-phosphotungstic acid hydrate (PW11) hybrid membranes (PEC/PW11) were prepared by blending sodium alginate (SA) and gelatin (GE), followed by incorporating with PW11, and then crosslinking by g-Glycidoxypropyltrimethoxysilane [26]. Swelling experiments showed that the degree of swelling of the PEC/PW11 hybrid membranes increased with increasing PW11 content or water content in feed. Sorption experiments demonstrated that when PW11 was no more than 9 wt%, both the sorption selectivity and diffusion selectivity increase with increasing PW11 content, then decrease with further increasing PW11 content.

In [27] blend and filled membrane from polyvinyl alcohol and sodium carboxy methyl cellulose were produced and used for separation of acetic acid– water mixtures over the concentration range of 81–98 wt.% acid in water including pure (100 wt.%) water and pure (100 wt.%) acetic acid. In the feed concentration range of around 2–19 wt.% water the unfilled membranes show sorption of around 5–40% while the filled membranes show around 10–55% sorption.

Poly (acrylonitrile-co-methyl acrylate) copolymer was used for making pervaporation membrane [28]. This membrane was used for separation of acetic acid–water mixtures over the concentration range of 80–99.5 wt.% acetic acid in water. Interaction parameters based on Flory–Huggins lattice model and engaged species induced clustering (ENSIC) model was used to explain swelling of the membranes.

Unfortunately, most article in the sphere of pervaporation concerning with development and testing of new membranes and only partial characterization of physicochemical properties including liquid sorption. At the same time it is obvious that development of mass transport theory is necessary for increasing of practical application of pervaporation. Experimental data presented in the paper demonstrates the significance role of swelling in pervaporation selectivity.

EXPERIMENTAL

Pervaporation experiments [29] were carried out on laboratory cell using nonporous membrane (thickness 70m) based on cellulose hydrate (Secon, FRG) at 30 ± 0.1°C. Liquids under separation were mixtures of acetic acid and water. The downstream pressure in the vacuum chamber was maintained at 1,

20, 30, or 50 mm Hg with an accuracy of ±0.1 mm Hg. The equilibrium sorption of membranes in mixtures was studied at 30 ± 0.1°C [29]. Membrane samples (0.6-0.7 g) were placed in thermostated aqueous-organic solutions. The weight change was determined gravimetrically with the error ± 10^{-4} g. The experiment was continued until equilibrium was attained (10-14 days). Further, free liquid was removed from membrane surface using filter paper and the sample was placed in a flask thermostated at 60°C. The flask was connected to a vacuum pump and trap cooled in liquid nitrogen where desorbed vapor was collected as condensate. The downstream pressure was 1 mm Hg. Accuracy of measurements were controlled by weight measurement before and after the desorption experiment. The liquid was completely removed from the membrane after 2 h of desorption. The compositions of liquid mixtures and vapor condensates were determined by gas chromatography (GC-2010PLUS device, Shimadzu, Japan).

Flux of pervaporation was defined as the amount of permeate collected in trap on unit area of membrane at unit time [1, 13].

Degree of swelling $S^{(E)}$ was obtained as percentage ratio of the weight of sorbed liquid in membrane to the weight of dry membrane.

RESULT AND DISCUSSION

The results obtained in pervaporation separation of water–acetic acid mixtures and sorption tests are presented in Figures 1-3.

Figure 1 demonstrates the dependence of water concentration in permeate on water concentration in feed at different downstream pressures of permeate. Vapor–liquid equilibrium curve is presented for comparison with pervaporation data. The direction of the permeate–feed concentration curves of pervaporation differs essentially from vapor–liquid equilibrium curve, and permeate is considerably enriched with water. As downstream pressures increases from 1 to 30 mm Hg for each compositions of feed mixtures, water concentration of permeate grows. As downstream pressures further increases (from 30 to 50 mm Hg), the composition of the permeate remains unchanged. Flux through membranes decreases and becomes as low as 27 $gm^{-2}\ h^{-1}$ at 30-50 mm Hg.

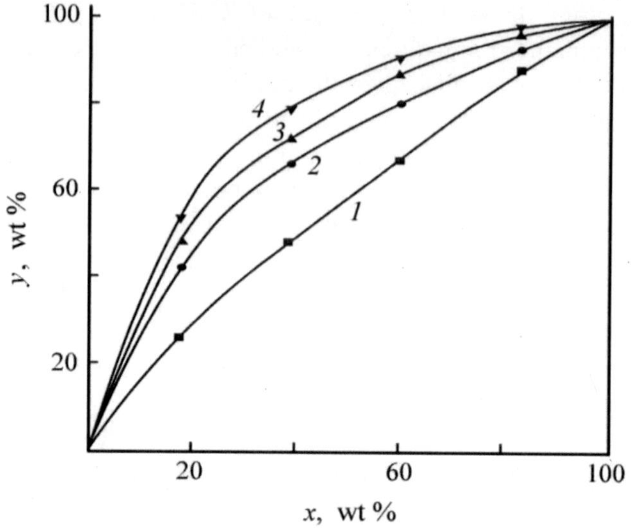

Figure 1. The dependence of water concentration y in permeate on water concentration x in feed in pervaporation of water–acetic acid mixtures at downstream pressures of (2) 1, (3) 10, and (4) 30 mm Hg at 30 °C. Line 1 corresponds to the vapor-liquid equilibrium.

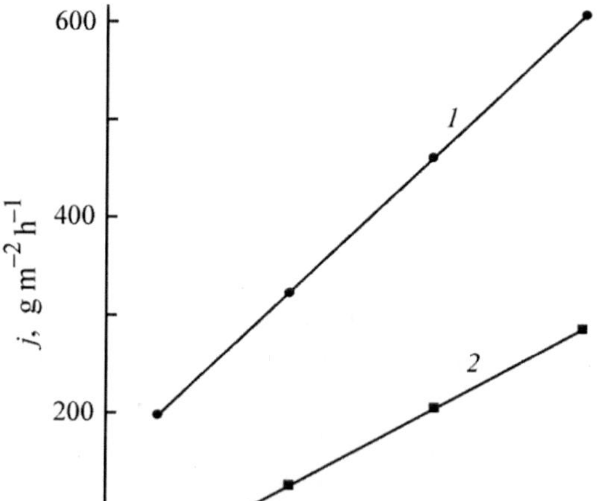

Figure 2. The dependence of flux j on water concentration in feed x in pervaporation of water–acetic acid mixtures at downstream pressures of (1) 1 and (2) 10 mm Hg.

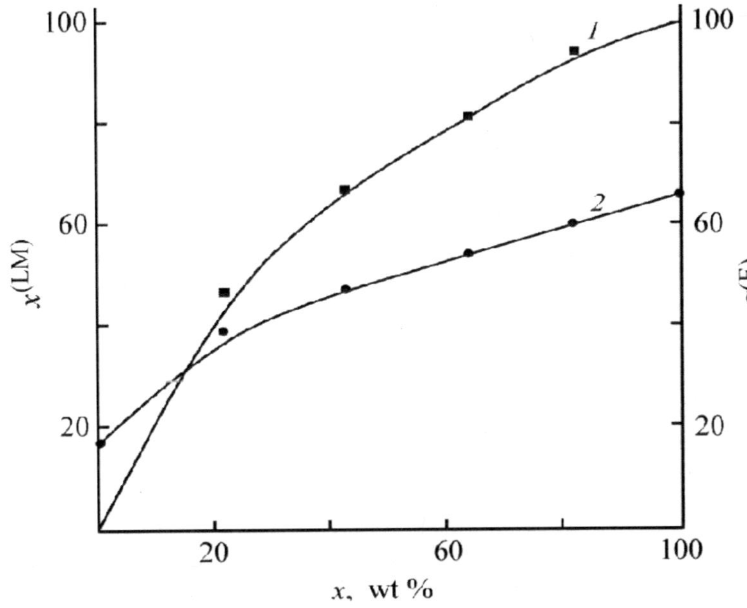

Figure 3. The dependence of (1) equilibrium water concentration $x^{(LM)}$ sorbed by membrane and (2) degree of swelling $S^{(E)}$ on water concentration in feed x.

It was assumed that in the case of low fluxes and composition independence of downstream pressure over 30 mm Hg evaporation through the membrane was close to the equilibrium state. The composition of the sorbed liquid in polymer was corresponded to the equilibrium data, which presented in Figure 3.

As equilibrium is attained, liquid sorbed by membranes is strongly enriched water compared with the starting mixtures (Figure 3). Effect of feed composition on the degree of swelling $S^{(E)}$ is also presented on Figure 3. As can be seen, the degree of swelling increases with the increase of water concentration up to 64wt.% and becomes linear at more than 20 wt.%.

Considering definition that factor characterizing liquid separation in the swollen layer of the membrane can be recovered from separation factor of pervaporation, it was found that pervaporation selectivity up to 60% determines by liquid sorption at downstream pressure over 30 mm Hg.

Conclusion

The sorption characteristics (equilibrium swelling and composition of sorbates) in pervaporation of water–acetic acid mixtures of various compositions were studied. It was shown that in pervaporation of these mixtures at downstream pressures of 1-50 mm Hg membranes demonstrate dehydration properties and permeate preferably water due to difference in the volatilities of acetic acid and water. As downstream pressure increased as flux and efficiency of separation decreased in all cases of separation because of reduction of driving force. At downstream pressure of 30 mm Hg and higher contribution of swelled membrane layer to the values of selectivity was significant and achieved up to 60%.

References

[1] Baker, R.W. *Membrane Technology and Applications*; second ed.; Wiley: New York, USA, 2004; pp. 355–392.

[2] Bolto, B.; Hoang, M.; Xie, Z. *Chem. Eng. and Proc.* 2011, *50*, 227–235.

[3] Koops, G.H.; Smolders, C.A. In Book *Pervaporation Membrane Separation Processes Estimation and evaluation of polymeric materials for pervaporation membranes*; Huang, R.Y.M.; Ed.; *Elsevier*: Amsterdam, KN, 1991; pp. 253–278.

[4] Aptel, P.; Challard, N.; Cuny, J.; Neel, J. *J. of Membr. Sci.* 1976, *1*, 271–287.

[5] Uragami, T.; Saito, T.; Takashi, M. *Carbohyd. Polym.* 2015, *120*, 1–6.

[6] Volkov, V.V. *Russ. Chem. Bull.* 1994, *43*, 187–198.

[7] Sae-Khow, O.; Mitra, S. J. *Chromatogr. A.* 2010, *1217*, 2736–2746.

[8] Feng, X.S.; Huang, R.Y.M. *Ind. Eng. Chem. Res.* 1997, *36*, 1048–1066.

[9] Wang, Q.; Li, N.; Bolto, B.; Hoang, M.; Xie, Z. *Desalination*, 2016, *387*, 46–60.

[10] Zhang, Q.G.; Liu, Q.L.; Jiang, Z.Y.; Chen, Y. *J. of Membr. Sci.* 2007, *287*, 237–245.

[11] Villaluenga, J.P. G.; Tabe-Mohammadi; A. *J. of Membr. Sci.* 2000, *169*, 159–174.

[12] Rautenbach, R.; Albrecht, R. *J. Membr. Sci.* 1980, *7*, 203–215.

[13] Mulder, M. *Basic Principles of Membrane Technology*; *Springer:* Dordrecht, KN, 1991; pp. 1-21.

[14] Lipnizki, F.; Tragardh, G. *Separ. Purif. Rev.* 2001, *30*, 4–125.
[15] Feng, X.; Huang, R.Y.M. *Ind. Eng. Chem. Res.* 1997, *36*, 1048–1066.
[16] Zharov, V.T.; Serafimov, L.A. *Physicochemical Foundations of Distillation and Rectification*; Khimiya: Leningrad, USSA, 1975; pp. 1-240.
[17] Jounqueres, A.; Clement, R.; Lochon, P. *Prog. Polym. Sci.* 2002, *27*, 1803–1877.
[18] Enneking, L.; Heintz, A.; Lichtenthaler, R.N. *J. Membr. Sci.* 1996, *115*, 161–170.
[19] Hauzer, J.; Reinhardt, G.A.; Stumm, F.; Heintz, A. *J. Membr. Sci.* 1989, *47*, 261–276.
[20] De Bo, J.; Van Langenhove, H.; De Keijser, J. *J. Membr. Sci.* 2002, *209*, 39–52.
[21] Storonkin, A.V., *Thermodynamics of Heterogeneous Systems*; parts 1 and 2; Len. Gos. Univ.: Leningrad, USSA, 1967; pp. 1-447.
[22] Nicolis, G.; Prigogine, I. *Self-organisation in Nonequilibrium Systems: From Dissipative Structures to Order Through Fluctuations*; Wiley: New York, USA, 1977; pp. 1-491.
[23] Sarkhel, R.; Bandyopadhyay, M.; Bhattacharyand Bhattacharya, P. *Sep. Pur. Techn.* 2003, *30*, 89–96.
[24] Zhang, H.; Wang, Y.; Bai, P.; Guo, X.; Ni, X. *J. of Chem. and Eng. Data.* 2016, *61*, 213–219.
[25] Julloka, N.; Deforche, T.; Luis, P.; Van der Bruggen, B. *Chem. Eng. Sci.* 2012, *78*, 14–20.
[26] Chen, J. H.; Zheng, J. Zh.; Liu, Q.L.; Guo, H. X.; Weng, W.; Li, Sh. X. *J. Membr. Sci.* 2013, *429*, 206–213.
[27] Das, P.; Ray, S.K. *Sep. and Pur. Tech.* 2013, *116*, 433–447.
[28] Kuila, S.B.; Ray, S. K. *Pol. Eng. and Sci.* 2013, *53*(5), 1073–1084.
[29] Kuznetsov, V. M.; Toikka, A. M.; Kuznetsov Yu. P. *Rus. J. of Appl. Chem.* 2007, 80(6), 904 - 908.

INDEX

A

B

C

G

H

I

J

K

L

M

N

O

P

Q

R

S

T

U

V

W

Y

Z